Business Continuity Planning

Business Continuity Planning

Increasing Workplace Resilience to Disasters

Brenda D. Phillips

Indiana University South Bend, South Bend, IN, United States

Mark Landahl

Emergency Manager, Rockville, Maryland, United States

Butterworth-Heinemann
An imprint of Elsevier

Butterworth-Heinemann is an imprint of Elsevier
The Boulevard, Langford Lane, Kidlington, Oxford OX5 1GB, United Kingdom
50 Hampshire Street, 5th Floor, Cambridge, MA 02139, United States

Notices
Knowledge and best practice in this field are constantly changing. As new research and experience broaden our understanding, changes in research methods, professional practices, or medical treatment may become necessary.

Practitioners and researchers must always rely on their own experience and knowledge in evaluating and using any information, methods, compounds, or experiments described herein. In using such information or methods they should be mindful of their own safety and the safety of others, including parties for whom they have a professional responsibility.

To the fullest extent of the law, neither the Publisher nor the authors, contributors, or editors, assume any liability for any injury and/or damage to persons or property as a matter of products liability, negligence or otherwise, or from any use or operation of any methods, products, instructions, or ideas contained in the material herein.

Library of Congress Cataloging-in-Publication Data
A catalog record for this book is available from the Library of Congress

British Library Cataloguing-in-Publication Data
A catalogue record for this book is available from the British Library

ISBN: 978-0-12-813844-1

For information on all Butterworth-Heinemann publications
visit our website at https://www.elsevier.com/books-and-journals

Publisher: Janco, Candice
Acquisitions Editor: Romer, Brian
Editorial Project Manager: Lawrence, Lindsay
Production Project Manager: Raviraj, Selvaraj
Cover Designer: Bilbow, Christian J.

Typeset by SPi Global, India

Working together
to grow libraries in
developing countries

www.elsevier.com • www.bookaid.org

Contents

About the authors

Brenda D. Phillips, Ph.D., is the Dean of Liberal Arts and Sciences and Professor of Sociology at Indiana University South Bend. Previously, she taught emergency management at Oklahoma State University and has served as a subject-matter expert, consultant, and volunteer for multiple agencies, communities, educational institutions, and voluntary organizations. She is the author of Disaster Recovery, Mennonite Disaster Service, and Qualitative Disaster Research, the co-author of Introduction to Emergency Management, and the co-editor of Social Vulnerability to Disaster. She has written numerous peer-reviewed journal articles in the discipline of emergency management and disaster science with direct experience in researching hurricanes, tornadoes, earthquakes, tsunamis, and hazardous materials accidents, much of which has been funded by the National Science Foundation. Dr. Phillips has been invited to assist or speak at disaster programs in the United States, Canada, Mexico, People's Republic of China, Costa Rica, New Zealand, Germany, Sweden, and Australia where she has promoted evidence-based best practices for community safety. Dr. Phillips firmly believes in the extension of faculty expertise through volunteer service. With over 30 years of experience in the field of emergency management education, Dr. Phillips has volunteered for local emergency management planning committees and voluntary organizations, especially for high risk populations. She has served as an unpaid reviewer of city and agency emergency management plans and assisted with planning around disaster-time domestic violence, safety for people with disabilities, and elderly evacuation. She has led business continuity planning at multiple academic institutions and businesses. Her most meaningful volunteer activities have helped to rescue animals and rebuild homes after disasters.

Mark Landahl, Ph.D., CEM® has more than 23 years of experience spanning the fields of law enforcement, emergency management, homeland security, and higher education. After retiring as the Homeland Security Commander for the Frederick County (MD) Sheriff's Office, he began service in his current position as the Emergency Manager for the City of Rockville (MD). In addition to professional practice, Dr. Landahl teaches at several universities as an adjunct professor. In addition, Dr. Landahl publishes regularly with work appearing in as contributions to several edited texts, the Journal of Homeland Security and Emergency Management, Homeland Security Affairs Journal, and the Journal of Urban Management. He also is the Co-Editor (with Dr. Tonya Thornton - George Mason University) of the forthcoming book Law Enforcement in Homeland Security and Emergency Management.

Preface

We finished this volume during the COVID-19 pandemic, when both of us became deeply embedded in business responses to and recovery from the significant impacts of this worldwide event.

Brenda, a Dean at a public university, managed a massive shift of personnel as faculty moved online, staff began to telecommute, and a new way of life emerged. Her work had already included developing hundreds of business continuity plans at multiple academic institutions. Brenda then "pivoted" (to use the phrase of the year) into offering business continuity planning across her local community for businesses, social service agencies, and educational partners. She cochaired the Working Group for Employee Welfare, Health and Safety, and served on the Restarting Committee for her campus. Previously, she was a Professor of Emergency Management and volunteered and consulted worldwide.

Mark, a recently retired Law Enforcement Commander, and the newly appointed Emergency Manager for a local jurisdiction in the Washington, DC Metro area also managed through the crisis. After just 4 months on the job in a newly created position, where he had just barely had time to identify gaps in exiting Emergency Operations and Continuity of Operations Plans, the COVID-19 crisis struck. In addition to the facts that plans were outdated, had not been exercised, and staff were not familiar with the contents, the EOP did not even consider public health emergencies. Despite these limitations, Mark led through the crisis with senior city leadership to take early steps to link with county public health officials, identify resource needs and move to early acquisition (ahead of all other regional jurisdictions), hold an immediate precrisis COOP tabletop exercise, shift services to remote operation, and maintain essential city services (including law enforcement and public works operations). Due to the work of senior city staff, despite facility closures, positive staff tests, and an unscripted response, the city never "closed."

The lessons learned from COVID-19 demonstrated both the value of planning and being flexible, adaptive, and resourceful. Empirical research supports the need for both planning and flexibility, as readers will uncover in this volume.

Throughout the chapters, we have tried to provide solid science coupled with the practical realities of how to implement evidence-based best practices. In addition, we have offered lessons learned from our own deep experiences with teaching, researching, and managing real events. We have tried to make this book user-friendly with guidance and inspiration in every chapter. We have kept small business in mind, because they have the highest failure rates in an emergency. We want them to succeed.

Acknowledgment and the Dedication

We would like to thank our families for their support during the writing of this volume, particularly David M. Neal, Frank and Mary Jane Phillips, Jesse, Sloopy and Scarlet Phillips, Charlotte Landahl, Casey Landahl, Abbey Landahl, Judy Landahl, and Marilyn Vettel. In addition, we also want to thank our team at Elsevier Press including Lindsay Lawrence, Ashwathi Aravindakshan, Katerina Zaleva, Srinivasan Bhaskaran, and Selvaraj Raviraj.

It is our fervent wish that this volume inspires businesses, agencies, and educational institutions to be more prepared for the "next one" whether it is a pandemic, natural disaster, terror attack, or another untoward event. Business continuity planning is possible at varying scales and levels of effort whether the business is a one-person operation or one that spans multiple continents.

We dedicate this volume to those who lost their lives in the COVID-19 pandemic, in hopes that we have helped to make the world a little safer through our work.

CHAPTER 1

What is business continuity planning?

Introduction

Disasters disrupt the normal, everyday routines and functioning of a business, organization, or agency (Quarantelli, Lagadec, & Boin, 2006, see also Quarantelli, 1998; Perry & Quarantelli, 2005). Whether a natural disaster, a hazardous materials contamination, a terror attack, a pandemic, or a cyberattack, each has the potential to stop a business from customary operations and cause the business to fail. Not surprisingly, when productivity ceases, economic losses occur as business customers, patients, or clients leave, and employees face the loss of a paycheck, benefits, and a job. Surrounding communities discover their tax base has been eroded. Businesses that rely on each other for parts and services, to generate customer traffic, and to create an interdependent economy may find such partnerships undermined by the disaster. Government programs may be hard-pressed to provide sufficient resources to businesses including physical assets, loans, and economic stimulus funding.

Business continuity planning (BCP) addresses what industries of all sizes and kinds need to think through to stem such losses, maintain customers, support employees, and continue producing goods and services for the surrounding community and business sector (Paton & Hill, 2006). Business continuity planning thus represents an intentional, thoughtful, and stepwise process that focuses on the dangers a business might face and what can be done to reduce losses and survive.

Conducting business continuity planning starts with understanding the kinds of risks that businesses might face. In the first part of this chapter, readers will learn about the kinds of area hazards that might imperil their business operations or partners they rely on. The latter part of the chapter then situates business continuity planning and outlines basic practices for launching such an effort. The goal is to provide general guidance that can be scaled up from small to large businesses, and cover a range of types of businesses, agencies, and operations. Throughout the volume, readers will find a range of examples to inspire

1

Business Continuity Planning. https://doi.org/10.1016/B978-0-12-813844-1.00009-9

planning in their setting, whether they own and operate a traditional business like a restaurant, an agency that serves people who are homeless or homebound, or a value-added community entity like a museum or university.

What most businesses do (not) do for a disaster

This volume should provide a red alert to the business community, in part because most businesses do the bare minimum to stay in operation (Orchiston, 2013; Webb, Tierney, & Dahlhamer, 2000, 2002). Other than a first aid kit and maybe a fire extinguisher, most businesses fail to prepare for significant losses and disruptions including efforts on how to adapt to a new normal (Webb et al., 2000, 2002). Restaurants may require training in the Heimlich maneuver, but they may not address what employees should do during a tornado, active aggressor event, or cyberattack. Smaller and micro-businesses seem to be particularly at risk in disasters given their levels of preparedness (Orchiston, 2013).

Beyond such basics, businesses largely fail to conduct any level of emergency planning (Orchiston, 2013; Webb et al., 2000). If they do, such planning is likely to be minimal such as a posted set of instructions on what to do if a tornado threatens. While some worksites engage in practicing response behaviors, like school and university fire drills or active aggressor training, most businesses do not undertake such essential actions. High rise building offices may include signage indicating which direction to evacuate, but rarely if ever practice exiting like Morgan Stanley's life-saving efforts did before September 11. Even more broadly, an economically valuable tourist area may fail to realize the full extent of the area's natural hazards, a reality that led to over 300,000 deaths in the 2004 Indian Ocean tsunami across 13 nations.

Businesses overall also fail to plan for where they will go if they need to relocate temporarily or permanently, which may determine their success or failure. Few businesses train their employees how to remotely access computers should a building become unavailable or a quarantine be imposed. Even fewer seem to identify ways to adapt should an emergency occur including even short-term work stoppages during an infectious disease outbreak. Businesses also lack planning in who can step in if a critical employee becomes injured or unavailable such as having to drive from a distance where they evacuated.

Businesses do pursue risk reduction along more traditional rates, such as buying insurance. Yet, businesses need to regularly assess insurance coverage as it may not include all area hazards, such as floods or high winds. Businesses may need hazard-specific policies, which can be cost-prohibitive especially for smaller businesses. Business interruption insurance may be available but might not cover all the kinds of losses experienced in a pandemic or may be unaffordable for small businesses. Rainy-day funds for many businesses come from personal savings, when a more stable source that covers specific expenses is what businesses need. Without business continuity planning, the ability of the business to fulfill its mission is at stake not to mention paying mortgages or rents, business loans, and payroll.

Stop for a moment and imagine that you are currently working in a one-floor building – look around at your workspace. If your company has a 3-foot flood, what would be lost? How would you continue working if you lost everything below 3 ft, which would probably include electrical outlets, computer/Internet connections, copiers or the motorized bases for assembly lines and production? What about 6 ft of water? Patient beds, office pods, and other critical resources would be lost. Or, consider area hazards besides a hypothetical flood. What natural, technological, or hazardous materials might impact your actual workplace? If your work relies on navigation, what disruptions might be anticipated from a solar flare or geomagnetic storm? Then, stop and think about disasters that might indirectly impact you, like a power outage or loss of utilities because the city's water treatment system became compromised. How would you get back to normal? Where would you work from while your place of work is restored?

Readers may also worry that they lack expertise in disaster management, such as having professional staff including emergency planners, environmental health experts, or risk managers. Lack of expertise can become a reason to not do something – but with business continuity planning (BCP), that is not necessarily the case. Businesses, schools, agencies, and faith settings can use available software or open source templates to launch and complete a BCP by using this volume. Area experts, such as local city and county emergency managers, can be asked for their advice. Universities may be willing to extend their faculty experts to support businesses. Thus, while a lack of expertise may seem intimidating, it can be overcome by collaborating or partnering with those who do hold such knowledge and by working through a stepwise process to produce a viable business continuity plan.

Even with a partner or expert, an additional problem can rise up: that of the lone planner or over-reliance on a consultant or contractor to create the plan. The lone planner approach is a mistake, because best practices for planning must involve a wide set of eyes and ears on the issues that need to be covered (Der Heide, 1996; Quarantelli, 1985). Without involving a local team, plans created by individuals or even by specialized consultants are likely to be insufficient or to include strategies that will not work or that people will not use. Schools and universities make this mistake when a high-level planning team writes the plan without consulting the stakeholders. Dealing with a quarantine might mean converting classes to online platforms, which makes sense unless the faculty lack the knowledge or computer resources to deliver online classes, like what happened during the COVID-19 pandemic. Some classes may be more difficult to offer via an Internet-based solution, for those who had not been brought into the planning and made ready. While classes like anatomy or dance can be offered successfully (Attardi, Barbeau, & Rogers, 2018), doing so requires some understanding about distance education platforms and pedagogies. Students also need to be ready and willing to maximize online learning opportunities to succeed (Langfield, Colthorpe, & Ainscough, 2017) and must have access to the hardware and connectivity that online learning requires. In short, everyone must be part of the planning effort or at least consulted and then advised and trained on the plan. Only then will an enterprise be ready to respond and recover with resilience.

What does seem to inspire action in planning for disasters is going through one. Following a flood or tornado, businesses are more likely to prepare (Marshall, Niehm, Sydnor, & Schrank, 2015). Such events, whether local or at a distance, can serve as inspirational and motivational moments to promote business continuity planning. The COVID-19 pandemic of 2020 has served as such a call to action, but pandemics are not the only hazard that businesses will need to address.

Hazards that become (business) disasters

The next section of this chapter scans a wide array of hazards that a business might face. Each hazard presents some unique challenges and not every hazard will become a disaster – again, defined by disrupting the normal operations of the business and its surrounding community. Even though each hazard might not seem to be present in the jurisdiction where readers live and work, it is worthwhile to review each hazard for the examples contained within. Each example has been chosen to raise awareness and motivate planning, from natural hazards through those made or caused by human actions.

Natural hazards

Natural hazards, by themselves, would do little harm to businesses. Rather, hazards become disasters occur because businesses locate their sites and people into places of significant risk (Quarantelli, 1998). It may be that business owners cannot make a choice, such as when and where people find employment in seasonal work like farming, fishing, or tourism. But sometimes businesses knowingly make avoidable choices, including building in an area of known earthquake activity like San Francisco, Tokyo, or Quito. To understand the potential impacts of natural hazards, this section will reveal some of the consequences when natural hazards and businesses fail to co-exist. Conversely, readers will discern the benefits of making informed choices to mitigate risks, particularly the practice of business continuity planning.

Tsunamis and cascading events

Naggapattinam Hospital, located in southeastern India, lies a few kilometers inland from the Indian Ocean. On December 26, 2004, a massive 30–40-foot tsunami pushed across shoreline villages and rushed toward the hospital. From the back of the sixty-building compound, a woman yelled out a one-word warning of "water, water." With 350 patients at risk, including those in neo-natal and intensive care units, medical staff and family members scrambled to carry patients to second floor levels. Water inundated the first levels of most buildings, leaving behind muddy debris and heavy damage. Survivors from the outlying area soon began to arrive, bringing the bodies of loved ones who could not be saved. At least 10,000 people died along a 5 km stretch of India's coast-line that day – but not a single patient perished at the hospital (Phillips, Neal, Wikle, Subanthore, & Hyrapiet, 2008).

Why did patients, staff, and family members survive in a situation where they had only themselves to respond? Just 1 month prior to the tsunami, hospital staff had trained on their plan, which determined how staff would handle an emergency. Staff mobilized quickly to restore operations and, with the help of outside funding, returned to serving as the region's only major hospital loca-tion. They not only saved their patients but tended to the injured and deceased the same day. Within a few months, in an area quite isolated from large city resources, the hospital returned to normal operations. As an added precaution, staff members moved critical hospital operations further inland in case another tsunami occurred.

Hurricanes and cyclones

Hurricane Maria slammed into Puerto Rico in 2017, devastating utilities, road-ways, and businesses of all sizes. The following year, the Federal Reserve Bank of New York conducted a survey, finding that 77% of businesses, the majority being small, suffered significant losses. Most of those businesses faced lost rev-enues coupled with increased expenses, damaged assets and property, and accu-mulating debt (Hamdani, Mills, Silva, & Battisto, 2018). Businesses also lacked sufficient insurance coverage to address losses and over one-third did not hold any insurance. Of those who did have insurance, only 6% indicated that their claims had been fully met (Hamdani et al., 2018).

One year earlier, cyclone Winston had torn through Fiji's islands. The worst such storm recorded in the region, Winston caused catastrophic damage to peo-ple, homes, businesses, schools, infrastructure, and hospitals. Businesses lost utilities, faced daunting travel on roadways, and dealt with an airport struggling to regain functionality. With 15% of the population newly homeless, businesses lacked employees to return to normal operations (Aquino, Wilkinson, Raftery, & Potangaroa, 2018). Damage reached $470 million or 10% of the nation's gross domestic product (Varandani, 2016). In addition to that economic impact, Fijians needed help to rebuild after catastrophic damage to schools and hospitals. Rebuilding stronger requires economic and political will to do so, which is more likely to occur after a disaster. As is common after such storms, Fiji opted to increase their post-disaster resilience with new construction codes (Aquino et al., 2018). Ideally, businesses want to anticipate such losses and to engage in efforts that reduce losses in advance but it can be challenging to garner the attention of those with the funds and means to do so until after something terrible happens.

Floods

Flood waters leave an ugly situation for businesses to deal with. Three general kinds of floods inundate businesses: storm surge (from hurricanes) and both river and flash flooding (e.g., from heavy precipitation or ice melt). Storm surge from a hurricane can be powerful depending on the strength of the hurricane that pushes water inland. In 2005, hurricane Katrina had just fallen to a category 4 strength (with 5 being the strongest category) as winds pushed a massive storm surge across multiple communities in Louisiana, Mississippi, and Alabama. The surge reached 40 ft high in some places, above tree and structure levels, and then overtopped, undermined, and blew out levees designed to protect homes and businesses. Many people and employees in the City of New Orleans evacuated the day before the storm surge arrived, but not in time to save over one thousand people. Businesses suffered terribly, as owners awaited the lengthy dewatering of the city before they could assess the damage in person. Even then, it would take years to restore utilities, clean up, commence repairs, and rebuild. Every business sector, in a city of over half a million people, sustained massive losses.

River flooding can sometimes be anticipated but can also happen very quickly depending on precipitation totals. In late 2018, the Aube River in France flooded, causing homes and bridges to collapse, and claiming lives when a normal 3-month rainfall total fell in one night (BBC, 2018). Heavy precipitation swelled rivers, flooded streets, and blocked roadways prompting rooftop evacuations from homes and businesses in an area appreciated for tourism opportunities. Reports indicated that over-building in an historic floodplain had contributed to the tragedy. The tourist sector became devastated, in a situation and scenario that could have been prevented with better mitigation and building codes.

Flash flooding typically happens even more quickly, and usually due to heavy rainfall in a specific location. A 2016 flood devastated a commercial sector of Ellicott City, Maryland in the U.S. Though some businesses rebuilt after the 2016 flood, others relocated when legislators opted to demolish damaged structures in a mitigation (meaning risk reduction) procedure called a buyout. Yet,

2 years later, 8 in. of rain fell in a short time span and flooded the same area. Once again, business owners and employees awaited rescue as cars floated down the street and then faced the daunting task of mucking out the muddy refuse. Defying the odds, two eerily similar 1000-year events happened within 24 months in the same site.

Floods leave an ugly situation behind to deal with from the water, mud, and debris that enters businesses. The water is not passive either, as it moves furniture around, seeps into insulation and electrical systems, destroys electronic equipment, and devastates machinery. Flooding also destroys landscaping and signage, undermining the elements that attract customers to the site. Items inside the business, particularly hazardous materials, can contaminate the business and the surrounding area. Structural elements, the building itself, and the non-structural contents can become total losses. People cannot return to work in such conditions and face the loss of their incomes. Having a plan to protect the structure and its contents, and a plan on how to regroup can matter in the recovery of an affected business and its human resources.

Earthquakes

Earthquakes create both seen and unseen damage, which can take a while to uncover. In Santa Cruz and Watsonville, California (U.S.), a 1989 earthquake slowly revealed underground utility damage and – a year later – enough structural harm that a school had to be closed. Several historic earthquakes reveal further both the risks and the potential to address risks. The Kobe, Japan earthquake of 1995 (Richter 6.9) claimed the lives of over 6000 people and damaged 50,000 buildings (Yamazaki, Nishimura, & Ueno, 1996). In addition to how the event harmed people, the event caused considerable disruption to businesses as the earthquake damaged bullet trains, highways, and utilities that took years to repair. Damage assessment revealed, though, that high rise buildings built since the nation imposed new seismic codes in 1981 survived well. As realtors like to say, "location matters" especially when selecting the location for a business. Being in a site with new seismic codes meant the difference between businesses re-opening or closing forever.

Furthermore, earthquakes occur worldwide, and significant global inequality means that not all places and people will fare as well. The 2015 Nepal earthquake, for example (Richter 7.8) killed close to 9000 people, left about 3.5 million homeless, and caused over 10 billion US dollars in economic losses in a significant hit to its gross domestic products (Goda et al., 2015). Nepal is also home to World Heritage Sites, which preserve archaeological and cultural history. Such sites represent employment opportunities to locals who support the heritage tourist industry. And, while tourism may seem trivial in the face of such loss of life, the industry represents a means to restore people's livelihoods so they can rebuild and care for their families. UNESCO thus launched an effort to restore the sites in the Kathmandu Valley (UNESCO, n.d., https://whc.unesco.org/en/news/1480/). Nepal's experience demonstrates the vulnerability not only of the area but how many businesses may need support financially from governmental and non-governmental sources to survive.

Volcanos

Indeed, tourism is an important industry. In 2018, a volcanic eruption prompted tourists to avoid Hawaii, causing tourist-related industries to face economic impacts while employers and their workers shouldered the loss of both homes and businesses. Volcanic eruptions, while seemingly not as common as other hazards, can cause significant disruptions even at a fair distance away. In 2010, the Eyjaafjallajökull, Iceland volcano eruption caused 6 days of air travel delays. Ash spread into northern Europe and airlines in 20 countries closed their airspace. The event affected 10 million travelers including business employees and tourists (Bye, 2011). Clearly, the airline industry suffered impacts but so did people trying to conduct business operations. The lesson learned in 2010? That natural hazards can cause both direct impacts to businesses such as closures requiring repairs and rebuilding as well as indirect impacts that undermine normal operations worldwide.

New Zealand is home to dozens of volcano threats. Potential impacts from an eruption could disrupt 1.4 million people which contribute to 35.3% of the nation's gross domestic product including areas important to wine production and tourism (McDonald, Smith, Kim, Cronin, & Proctor, 2017). Land use planning has contributed to creating concentrated areas of business activity, such as an industrial or manufacturing center. When co-located with volcanic risks, and without sufficient disaster planning, the economic losses could cost billions in local currency depending on the magnitude and scope of an event (McDonald et al., 2017). The Mount Zao area of Japan has also experienced volcanic activity that could unduly affect area industry. Yet a survey found that business owners remained unaware of risks and did not trust pre-eruption warning messages (Donovan, Suppasri, Kuri, & Torayashiki, 2018). Understandably, volcanic eruptions and earthquakes remain difficult to predict and warnings may not result in an event. Similarly, tornadoes usually carry a very short warning period and can cause a direct hit or completely miss the building next to one that is demolished. Such seeming unpredictability can make it challenging to know whether to plan for business disruptions. Actually, the answer is to do so anyway – because disasters happen all the time and businesses will be glad they had a plan.

Tornadoes

Sometimes the magnitude of an event does not matter as much as the impacts. Meteorologists rank tornadoes, which occur most often in the United States, on the Enhanced Fujita scale from EF1 to EF5, with EF5 being the strongest. Yet, an EF1 or EF2 can damage or destroy a manufactured home, a mobile van used for businesses, or a barn. Or, a tornado could entirely miss homes and offices but tear up an agricultural field, injure livestock, sink harbored boats, or toss parked airplanes. It is where the tornado hits, the kind of structure it impacts, and what is inside that determines the level of destruction and its impact on area businesses.

Tornadoes give us an opportunity to think about the range of businesses that such vortices could damage. Consider, for example, small businesses inside people's homes or vehicles or the workspaces of telecommuters. A home could serve as a site for childcare, insurance providers, accountants, carpenters, or caterers. A vehicle might function as a mobile business for electricians or plumbers. If the event is large enough, such as the EF5 that devastated Joplin, Missouri in 2011, nursing homes may be torn apart or the hospital may become unusable. A direct hit on a business can leave it as a complete loss, while an indirect impact may cause loss of utilities, supply chains, or workers to keep it operational. Secondary and tertiary markets may suffer as well, when people cannot use their usual routes for gas, groceries, laundry, and related services.

Fairly simple and relatively inexpensive procedures can make a difference with tornadoes, such as a weather radio that provides geographically specific warnings. Training employees on how to respond to a warning message and the importance of doing so is free and can save lives so they can help to clean up and keep the business going. Designating best shelter locations also represents an easy step, which involves identifying a space where employees and customers can put as many walls between themselves and the outside as possible or go underground. Scaling up the protection and cost, another option might include installing a safe room, which can cost several thousand dollars (USD) but save lives. Beyond life-saving efforts for employees, the business will need to consider how it will recover. Locations that serve the injured, such as health care settings, will need to develop far more detailed business continuity plans including the ability to relocate, add staff, communicate internally and externally, and offer mobile hospital units to supplement health care operations (Kearns et al., 2017). Businesses impacted by the tornado will need to think through where they will go if displaced and how to become operational again – that is the purpose of a business continuity plan.

Hazardous materials

It is wise to think about hazardous materials in two ways: as potential hazards themselves or as the result of another event. Tornadoes, for example, can carry hazardous materials aloft and, by carrying them in a debris ball, leave them strewn hundreds of miles away. Floods also move materials like pesticides, gas, and oil into new locations. Thus, by themselves, potentially hazardous materials can be used safely but when a natural hazard arrives such materials can become problematic.

Hazardous materials can also cause disasters when things go wrong. That was the case in 2015 when 30 tons of ammonium nitrate exploded in West, Texas. Fifteen people died, including 3 residents and 12 first responders. Significant damage happened to area homes and businesses, a nursing home, and a school, with debris extending for several miles beyond the plant (Laboureur et al., 2016). In 2019, another plant-related explosion took the lives of 47 Chinese and injured many more at a chemical industrial park in Jiangsu Province. The event caused a massive evacuation of thousands of people in an area critiqued previously for a lack of safety protocols (Ramzy & Bradsher, 2019). Even a temporary displacement of workers meant disruptions to unaffected businesses who lost their employees for a time. Business continuity planning readies a business for such personnel disruptions and lays out how to restart and reopen.

Technological disruptions and hazards

In the last 20 years, we have come to think of technological disruptions in several ways. Cyberattacks have increased, ranging from denials of service to ransomware and massive phishing schemes to capture critical data. The consequences for businesses can be significant, from downtime to loss of records and potential loss of the business. Cybercriminals targeted hospitals worldwide in 2018, causing disruptions to services along with significant financial impacts. Hospitals in Ohio and West Virginia (states in the U.S.) had to divert ambulances away from their emergency rooms when a ransomware attack compromised abilities to deliver health care (Mathews, 2018). Though the impacts were short-lived, the event prompted attention to these new threats to business continuity operations. The cyber intrusions during the 2016 U.S. Presidential elections also created not only extensive federal investigations but impacts (and opportunities) to state and local elections systems, supporting businesses, and consultants who worked toward a more secure 2020 election.

Space weather has also emerged as a hazard that can cause technological disruptions that affect business operations. Though technically a natural event, geomagnetic storms and similar events have disrupted GPS and navigational technologies for airline and waterborne industries, impacted cellular services, and raised concerns about power grid stability. A significant solar flare occurred in 2017, undermining the positioning accuracy of satellites (Berdermann et al., 2018). Solar flare effects could also affect the navigational abilities of airlines and waterborne shipping industries. First noticed in 1859 during the Carrington Event, a geomagnetic storm caused problems with telegraph services and resulted in one death (Henderson et al., 2017). A similar storm today might prompt a "cascading event" which means that one event causes another. For example, a geomagnetic storm might disrupt communications used by ports and militaries or affect power grids that keep other utilities operational (Pescaroli, Turner, Gould, Alexander, & Wicks, 2017). Concern exists that, because of aging infrastructure, a storm like Carrington could take out the electrical grid on the U.S. eastern coastline. If such a storm happened during the winter months, consequences could be severe for people and businesses. Fatalities would occur in hospitals and nursing homes without heat for patients, or electricity to power durable medical equipment, emergency rooms, and trauma centers. Economic disruptions could total significant amounts during extended power outages, potentially reaching billions of dollars (Barnes & Van Dyke, 1990). Yet, most businesses fail to think about space weather although such events continue to happen or to set up an alternative means to power critical functions.

Terrorism and active attackers

Both domestic and international terrorism disrupt workplaces, schools and universities, ports, and more. As the 2017 mass shootings in Las Vegas and the 2019 New Zealand mosque attack showed, even recreational venues and worship sites do not remain immune from violence. No business, agency, or organization can ignore the potential of disruptive and deadly events, even from internal violence caused by disgruntled employees or from a family member who commits domestic violence at an office or factory.

Mass shootings, attacks of violence, cyber-intrusions, ransomware and the like all disrupt normal operations and require business continuity planning. Because such planning focuses on the aftermath of such an event, the intent of the planning should be to resume normal operations as soon as possible. However, both physical and psychological trauma will have occurred and should be addressed prior to or as part of resuming normal business operations. Companies may also need to activate their employee assistance programs to offer counseling and to have human resources assist workers with health care benefits and leave to address physical, psychological, and spiritual harm.

Yet, the September 11, 2001 attacks on the World Trade Center in New York City demonstrated that preparedness and training can make a difference. Though significant loss of life occurred among employees and first responders, thousands of people survived the event. They did so because they paid attention to a prior attack in 1993. Consequently, companies like Morgan Stanley engaged in a concerted effort to plan for another event and to train their people how to respond. In 1993, it took 4 h for their employees to evacuate down

60 flights of stairs. In 2001, it took 45 min though the company did lose 11 employees (FEMA, n.d.). Planning saves both lives and livelihoods. Re-opening after such an attack can be difficult emotionally and financially. To re-open, businesses may opt to hold a ceremonial event. Following a terror attack on London Bridge in 2017, Borough Market reopened with a moment of silence. An attack on a hotel in Mumbai, India in 2008 required three weeks of repairs to resume business and nearly a year to repair the most heavily damaged sections. Business owners re-opened following prayer services. Business continuity planning covers how to address the human impacts of terrorism and the challenges associated with re-opening after such a devastating event.

Pandemics

The influenza pandemic of 1918 killed as many as 50 million people worldwide in three waves that played out over nearly 2 years (Niall, Johnson, & Mueller, 2002). After COVID-19, it is not as hard to imagine such an impact, including that people did not go to work to reduce the spread of the disease. Today, due to the reduced ineffectiveness of antibiotics and the rapid, potentially global transmission of disease, businesses should prepare for interruptions due to pandemics.

Even small-scale events can be disrupting such as the efforts to contain Ebola in the U.S. and Europe in 2014. Though the number of people who became ill from Ebola outside of the African continent remained small, concern erupted over the potential dissemination of the highly fatal disease via airline travel. Though efforts quarantined many symptomatic travelers, some did use air transport including one who came to the United States and subsequently died. Confidence in air travel dropped, especially to Liberia and Sierra Leone, resulting in multiple airlines cancelling flights into or out of affected countries (Amankwah-Amoah, 2016). Given widespread reliance on the airline industry across Africa, significant costs occurred for tourism as well. As the disease declined over time, airline industries faced a difficult and protracted battle to restore passenger confidence. Airline travel also led to the 2003 SARS outbreak (Knobler et al., 2004). Ultimately, SARS affected 30 countries within 6 months including significant impacts in Hong Kong, Singapore, Vietnam, and Canada. Something as straightforward as early recognition and careful hygiene could have made a difference. For this reason, influenza immunizations and hand-washing often represent a frontline defense easily implemented by businesses everywhere, especially those with customer service missions and high social contacts including schools and universities.

Could an outbreak of an infectious disease cause a quarantine where you work or attend school? The entire world found out this could happen in 2020 when COVID-19 killed hundreds of thousands of people in an event still underway when this book was published. Business continuity planning should address such potential events to maintain or adapt business operations to meet potentially catastrophic economic consequences, including measures to protect the health of their workforce and the broader community. COVID-19 made this scenario very real to everyone, a situation that will be returned to throughout this volume. Appropriately, the remainder of this chapter walks readers through the types of disaster planning that can make a difference for businesses across a range of disasters.

Types of disaster planning

To prepare for disruptions and to avoid significant losses, businesses should engage in several kinds of planning efforts. The most common types of planning address mitigation, preparedness, response, and recovery. To distinguish between the planning types, consider the time frame that each focus on (see Fig. 1.1). The remainder of this chapter gives a brief overview of these types of disaster planning and where business continuity planning fits. Future chapters will focus in detail on how to create a basis for and write up a business continuity plan.

Mitigation planning

Mitigation planning ideally takes place before an event happens, with an eye toward reducing the potential impacts to human lives (employees and customers or clients), properties, inventory, and business functions. Mitigation

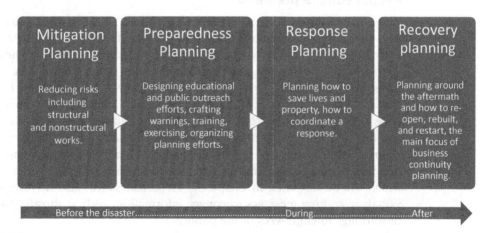

FIGURE 1.1
Types of planning, associated tasks, and focus of the planning.

planning usually begins with hazard identification and risk assessment, two tasks that will be addressed in Chapter 3 due to their relevance for business continuity planning. During mitigation planning activities, an emergency manager and planning team members (often from across the community) will review area hazards and determine their frequency and prior impacts.

Based on what they find, they will identify related risks and then prioritize them for actions that involve choices about structural and non-structural mitigation actions. As examples, structural mitigation might include placing barricades like concrete posts in front of a business to prevent vehicle entry from a poor driver or a terrorist attack. Or, a community might build a levee system to protect the downtown from regular river or flash flooding. At the business level, an enterprise might erect a floodwall, like Our Lady of Lourdes hospital did in Albany, New York. Damage from a flood had caused $20 million in losses during a 2006 event. In 2011, though, the water only reached the parking lot because of the floodwall.

Non-structural mitigation measures include actions like buying insurance to offset losses or comply with building codes. A 7.0 Richter magnitude earthquake in Christchurch, New Zealand occurred in 2010. Though costs from the earthquake reached significant totals, building codes resulted in no fatalities for this event (Bradford, 2010). In addition, the nation's Earthquake Commission provided coverage to homeowners that allowed them to rebound from some of their financial losses. Business continuity planning processes may reveal some actions that can be taken to mitigate or reduce losses before an event, in which a business may need to invest.

Preparedness planning

Sadly, an earthquake struck Christchurch again in 2011, killing 182 and heavily damaging the downtown business sector. Despite the massive destruction and thousands of injured survivors, health care facilities continued to provide high levels of care. Pre-disaster planning, coupled with well-trained teams, helped when loss of electricity and communications affected the hospital (Ardagh et al., 2012). Thus, the second kind of planning – preparedness – includes a wide range of activities from writing plans to training people on the plans and then practicing what to do. Practicing will usually include training workshops or conferences, tabletop or full field exercises to test the plan, and then revisions of the plan to address deficiencies uncovered during the exercises, which will be discussed in Chapter 7. Preparedness will also involve professionals in crafting messages to convey information, such as threats, watches, or warnings to employees or the public.

Company and public education efforts also develop during planning such as explaining how people will be warned and practicing how and where to evacuate. Businesses might put first aid kits into place including AED units and then train their personnel on how to use them. Restaurants might train staff on how to use the Heimlich maneuver if a customer chokes on food. A school or university would organize tornado or active attacker training followed by periodic drills to assess readiness. Social service agencies might do assessments of clients who are home-bound to see if they are ready for winter weather and the potential power outages and isolation that snowstorms can produce. Businesses would also want to create and set aside "go kits" to use from a distance, including what would be needed for telecommuting. In general, preparedness means planning and getting ready for something to happen by doing all you can in advance to think through and make sure your employees and business can face a crisis.

Response planning

Ideally, then, and given adequate mitigation measures and preparedness, the response period will go more smoothly with fewer negative impacts. Indeed, it is true that in many events, the "first responders" are fellow employees, neighbors, and family members. In the situation of an active attacker, most of the deaths and injuries will occur before the police can realistically arrive. For exactly this reason, workplaces may want to conduct active attacker training or offer first aid classes to enable better response in an emergency.

Most businesses fail to conduct anything significant in terms of response planning, with exceptions for health care facilities or businesses that have their own response units, such as a large industrial plant. Nonetheless, with only minimal training, employees will likely step in and care for each other. In 1978, a massive fire broke out at the Beverley Hills Supper Club near Cincinnati, Ohio. Waiters and kitchen staff turned into evacuation leaders, saving countless lives. Staff then provided building information to emergency responders as they tried to rescue customers and staff (Johnston & Johnson, 1989). From the Beverley Hills fire to the terror attack of 2001 to the Las Vegas shooting of 2018 researchers have learned that people can be relied on to respond well even with a minimum of training. Imagine what could happen with even more planning and training. That is the goal of emergency response time planning, to ensure that response goes as effectively as possible to save lives and property.

Recovery planning

The recovery period can linger from days to years, depending on the impact and the abilities and assets of an enterprise, agency, or organization to rebound and recover. Having a plan to recover from a disaster matters, because it can mean the difference between surviving financially or going under. Surprisingly, though, recovery planning remains limited in businesses as well as across the emergency management sector (Phillips, 2015). Recovery planning tries to address the steps and stages that a business or agency will take to deal with the consequences of an event. Ideally, a recovery – in the present case, a business continuity/recovery – plan will lay out what must be done to regroup, who will do that work, how it will be done, and with what assets. The plan outlines essential actions to take so that the mission of the business, agency, or organization continues, such as the delivery of health care and social services or the continued production of goods and services.

Recovery planning for businesses focuses on continuity of operations, starting with critical functions focused on the core of the business or agency – actions and efforts that must be completed at the most fundamental level. For a social service agency, that might be the continued delivery of nutritional resources to families in need or maintaining the availability of a domestic violence shelter. A bank would be concerned with transactions for customers both in person and electronically. A restaurant would want to continue producing meals to serve customers and retain employees. A government agency might need to determine how to telecommute or staff essential services while still supporting core public service needs. Business continuity planning focuses and leverages the enterprises' assets – especially its people – to stay afloat and maintain operations.

The purpose of this book, then, is to look beyond the time frame for mitigation, preparedness, and response planning into what happens as the smoke clears or the flood waters recede. How do employees address critical functions so that businesses can return to normal? What priorities should be pre-identified for action and along what kind of a timeline? How can businesses manage without key staff or adequate inventory? What kinds of losses should be planned for to safeguard profits and remain viable? Where should the company relocate if necessary? How long can the company go without restoring operations and with what kinds of impacts to the business and its employees? What kinds of scenarios should be considered given area hazards and contemporary threats? This book, while drawing on and providing mitigation, preparedness, and response planning insights, will be situated squarely in the post-event period as a function of recovery planning.

The benefits of business continuity planning

The best way to secure support for conducting business continuity planning is to inspire business owners to buy-in to the benefits which include:

- *Mission fulfillment.* Businesses exist for a variety of purposes, from making a profit to providing a service. The continuum of missions that they pursue represents the heart of business continuity planning. Places of worship will want to continue serving the people who attend and the faith traditions they follow. Schools and universities will want to keep students enrolled and moving toward graduation as well as pay their employees. Health care settings will want to keep taking care of patients and nonprofits will still want to care for clients. Corporations will need to keep producing and profiting from their products and stay in operation for years ahead. The prime focus of BCP is this: to keep doing what you have been doing that meets a need and generates results. Mission statements, which serve as the heart of any business or strategic plan, also drive business continuity planning.

- *The financial bottom line.* Perhaps the single most important benefit, from the eyes of the business owner, is the cost of restoring operations and getting back to normal. Business continuity planning can reduce the time it takes to do so by identifying the potential losses that can be tolerated and setting up a means to serve the company mission, identify how to adapt, or when to close forever. Business continuity planning provides the means to sustain a business and emerge from a disaster resiliently.

- *Loss containment.* The BCP process requires a close examination of potential losses vis-a-vis area hazards and threats. A simple inventory of computers, for example, can reveal the cost of replacement should they become damaged. Perhaps insurance coverage could be adjusted to help with such a post-disaster expense. Moving beyond inventory to the costs of structural damage can raise eyebrows even more. Buildings can be very expensive to replace after an earthquake, including making them more disaster resistant. Similarly, a cyclone can wreak havoc on a harbor or airport or the industries that rely on such sites. A straightforward inventory of content, building, and asset losses given the most likely scenarios in the area can reveal the potentially catastrophic financial impact. A range of mitigation measures may need to be considered as insurance does not always cover all costs such as removing a foundation from a tornado, cleaning bio-hazards from a terror attack, or paying ransomware costs from a cybercrime. Business interruption insurance may be expensive or not available for all hazards. Realistically assessing potential losses can prompt business continuity planning for loss reduction.

- *Customer bases.* Businesses, agencies, and organizations rely on clients, customers, patients, students, and worship-goers to survive and need them to come back after a disaster. Yet, restoring customer traffic can be challenging. After the 2004 tsunami, the business sector had to rebuild completely in Vailankanni, India as did the fishing industry in Sri Lanka and the tourist sector in Thailand. Cancellations after the 2018 volcanic eruption in Hawaii reduced profits for companies reliant on travel vacations. The central business district in Seattle, Washington (U.S.) had to lure customers back to its historic district with a variety of creative means (Chang & Falit-Baiamonte, 2002). Getting the customer base back becomes an urgent action to take before businesses fail. Business continuity planning recognizes this challenge and walks businesses through ways to retain their valuable set of customers.

- *Human resource investment.* Businesses of all sizes rely on the people who work there. Because businesses invest in their human resources, they carry an ethical responsibility to safeguard not only the people but the costs of time and training they provided. The COVID-19 outbreak in 2020 revealed the potentially catastrophic impacts that people will experience. In the U.S. alone, over 30 million people lost their jobs leading to the largest unemployment numbers since the Great Depression of the 1930s. People drive businesses through their expertise, energy, and dedication to its mission. By looking at the impacts of a disaster on human resources (see Chapter 6), businesses can be ready to ride out personnel disruptions, adapt as needed, and continue contributing to the economy.

- *Diversity in the business sector.* Studies reveal that minority and woman-owned businesses fail more frequently after a disaster (Marshall et al., 2015; Webb et al., 2000, 2002). Business failure happens because women and people of color tend to have smaller businesses with fewer resources to recover due to historic patterns of marginalization and economic discrimination. Depleting their resources quickly, they experience higher risks to their survival. By engaging in BCP, such businesses can enhance their potential to endure, earn income, and continue to serve their local economy and community. The loss of such diversity in the community also reduces the scope and depth of services. Smaller businesses often cater to needs within localized areas, with a subsequent impact to those in need of such services: bookstores, coffee shops, childcare, bakeries, retail shops, financial planning, grocery stores and restaurants, and so much more. Also important, the multicultural economic base of a community can be harmed.

The benefits of BCP thus include not only staying in business but being able to serve a mission, support employees, and provide products and services to the broader community. While business continuity planning takes time, the investment is worth it. To launch your BCP effort, we next turn to some essential planning principles then move into chapters that walk planners, and their teams, through the planning process.

Essential principles for business continuity planning

The remainder of this book shows planners how to write, test, and implement a business continuity plan. Throughout this book, readers are advised to keep several principles in mind (based in part on Quarantelli, 1985):

1. *Planning requires involvement.* Business continuity planning should avoid the lone planner syndrome. Teams of people in the actual business should be involved in crafting a plan by integrating their knowledge and insights to yield a more effective final product. Consultants can be helpful in guiding the BCP process but should never write it for a business. Employers and employees need to make decisions about plan content and need to know the plan to be able to implement the plan. Employees cannot enact a plan if they pull it off a shelf and read it after the earthquake – assuming they can even find it in the debris or on a computer damaged by a cyberattack. Chapter 2 will address the importance of planning teams that involve a range of employees.

2. *Planning involves everyone.* Planning teams should also consider an array of planners and not just the experts. People with specialized expertise, like the information technology team or intensive care nurses, should be involved as they will know more about the planning context than others. And, while the CEO should be involved and approve the plan – so should the janitors who will know about security holes or the places that contain hazardous materials. Readers will find examples throughout this volume that emphasize involving a diverse set of planning team members.

3. *Planning is a process.* A process necessarily requires steps or stages that planners should move through. The next few chapters will provide instruction on how planners can walk their teams through business continuity planning steps. For example, businesses should always engage in hazard identification and an associated risk assessment to know what situations and threats they may face. Such a hazard analysis will become the heart of the business continuity plan and ensure that planning prepares a business for realistic future situations. Loss estimation will also represent an essential step, where planners engage in an inventory of hazards could impact and how much loss can be tolerated vis-a-vis potential loss reduction costs. Risk assessment and decision making around local hazards can be found in Chapter 3.

4. *Planning should be evidence-based.* All plans should always be based on evidence, on research, that supports planning assumptions. One myth, for example, is that role abandonment will occur in a disaster when people do not show up for work (Drabek, 2012; Johnston & Johnson, 1989). Role abandonment was a concern with some epidemics, for example, and for terror attacks. But even with the deadly Ebola epidemic, health care providers went to considerable efforts to serve fellow human beings. The same occurred during the COVID-19 pandemic when health care workers showed up at significant personal risk – and hundreds died as did newly designated essential workers including grocery store clerks, delivery personnel, and custodians (see Photo 1.1). Similarly, panic is largely a myth. During and after every single terror attack or active aggressor, people have helped each other. Even during the Las Vegas

PHOTO 1.1
Essential workers required personal protective equipment during the coronavirus pandemic to keep the workforce healthy and continue business operations.

shootings, where 58 died, people put their own lives on the line to save complete strangers. We can rely on each other during difficult times because role abandonment and panic are myths (Gantt & Gantt, 2012; Johnston & Johnson, 1989; Trainor & Barsky, 2011).

5. *A culture of continuity.* To ensure that employees take the planning effort seriously, businesses should initiate efforts that promote continuity as a routine matter (Alesi, 2008). Businesses can take advantage of seasonal hazards, such as the start of tornado or cyclone season to inspire attention and participation. Unexpected events, whether local or at a distance, can also be used. A cyberattack elsewhere, for example, can be used to remind employees of the importance of computer security procedures. The key to creating a culture of continuity is for management and business leaders to emphasize its importance to their employees verbally, in signage and messaging, and certainly by involving them in team planning efforts. When the organization believes in continuity planning, the employees will buy in as well.

Planning team size and plans

Before going much further, the size of the planning team and business can and should be considered as well. Smaller businesses and agencies with similar interests may want to band together to compare notes and encourage each other. They could convene a joint planning session to share ideas and information, because working together will surface more ideas and even spur collaboration and mutual aid agreements in a future emergency. Smaller businesses may also be able to complete their plans faster than larger businesses but should complete the steps outlined in this volume.

At the other end of the spectrum, very large businesses should consider having multiple plans and planning teams. For example, a university system may want each campus to have its own plan and to have plans within that campus including facilities, academic affairs, and human resources. Specialized units, like a health care center, would also need a very specific plan suitable to the disaster. A large industry might need separate business continuity plans for assembly line production, marketing, sales, and distribution. Such an approach can result in multiple plans that will require regular updating because employees may leave or move into new positions, the vendors or policies involved may change, and events may warrant fresh consideration. In such situations, it would be wise to task a specific individual with shepherding multiple plans to completion. Each unit involved that creates a plan will need to have a separate planning team and effort so that they can focus. The larger business will also want to create an umbrella plan that incorporates and considers what planning units have proposed – because a unit that expects to "buy 100 new mobile devices" for telecommuting would need consideration and approval from budget managers and IT – every plan needs to be linked and thought through, an effort that will be discussed further in Chapter 7.

Upcoming chapters

The remainder of this book provides readers with a user-friendly, evidence-based guide to business continuity planning. Chapter 2, *Setting the Stage*, will help planners to secure commitment from within the business and to build a planning team. Several key ideas will be explained based on the kinds of disruptions that hazards can cause, as described in this chapter. Chapter 3, *Pre-Planning Steps to Launch Business Continuity Planning*, will help planners and their teams to understand local hazards and to assess risks. Knowing about area hazards and impacts will be crucial to securing the attention of stakeholders and to designing the right kinds of continuity practices. Chapter 4, *Parts of a Business Continuity Plan*, will provide an overview of the typical components of a written plan including critical functions, work teams, asset inventories, coping strategies, and essential partnerships. In Chapter 5, *Planning for Disruptions*, readers will understand how to address infrastructure disruptions, downtime and displacement, and returning to operational status. Chapter 6, *Managing Human Resources*, looks at what happens to employees in a disaster and how businesses can support their people assets. Chapter 7, *Strengthening and Testing the Business Continuity Plan*, lays out training and exercise protocols to ensure that people know the plan and can implement their assignments. In Chapter 8, *Becoming More Resilient*, readers will think about how to recover and to reduce future disasters. Finally, several appendices will provide additional information and useful tools for planners and their teams. You can do this.

References

Alesi, P. (2008). Building enterprise-wide resilience by integrating business continuity capability into day-to-day business culture and technology. *Journal of Business Continuity and Business Planning, 3*, 214–220.

Amankwah-Amoah, J. (2016). Ebola and global airline business: An integrated framework for companies' responses to adverse environmental shock. *Thunderbird International Business Review, 58* (5), 385–397.

Aquino, D., Wilkinson, S., Raftery, G., & Potangaroa, R. (2018). Building back towards storm-resilient housing: Lessons from Fiji's cyclone Winston experience. *International Journal of Disaster Risk Reduction, 33*, 355–364.

Ardagh, M. W. , et al. (2012). The initial health-system response to the earthquake in Christchurch, New Zealand, in February, 2011. *The Lancet, 379*(9831), 2109–2115.

Attardi, S., Barbeau, M., & Rogers, K. (2018). Improving online interactions: Lessons from an online anatomy course with a laboratory for undergraduate students. *Anatomical Sciences Education, 11* (6), 592–604.

Barnes, P. R., & Van Dyke, J. W. (1990). Economic consequences of geomagnetic storms (a summary). *IEEE Power Engineering Review, 10*(11), 3–4.

BBC (2018). *France weather: Red alert as flash floods kill 10 in South-West. Available at(2018). https:// www.bbc.com/news/world-europe-45861428. (Last Accessed July 3, 2019).*

Berdermann, J., Kriegel, M., Banys, D., Heymann, F., Hoque, M., Wilken, V., … Jakowski, J. (2018). Ionospheric response to the X9.3 flare on 6 September 2017 and its implication for navigation services over Europe. *Space Weather, 16*(10), 1604–1615.

Bradford, M. (2010). Strict building codes limit New Zealand quake damage. *Business Insurance,.* Available at(2010). https://www.businessinsurance.com/article/00010101/STORY/ 309129982/Strict-building-codes-limit-New-Zealand-quake-damage. (Last Accessed June 14, 2019).

Bye, B. (2011). Volcanic eruptions: Science and risk management. *Science, 2.0,* . Available at(2011). https://www.science20.com/planetbye/volcanic_eruptions_science_and_risk_management-79456. (Last Accessed May 19, 2019).

Chang, S. E., & Falit-Baiamonte, A. (2002). Disaster vulnerability of businesses in the 2001 Nisqually earthquake. *Global Environmental Change, Part B: Environmental Hazards, 4*(2), 59–71.

Der Heide, E. A. (1996). Disaster planning, part II: Disaster problems, issues, and challenges identified in the research literature. *Emergency Medicine Clinics, 14*(2), 453–480.

Donovan, A., Suppasri, A., Kuri, M., & Torayashiki, T. (2018). The complex consequences of volcanic warnings: Trust, risk perception and experiences of businesses near Mount Zao following the 2015 unrest period. *International Journal of Disaster Risk Reduction, 27*, 57–67.

Drabek, T. E. (2012). *Human system responses to disaster: An inventory of sociological findings.* Springer Science & Business Media.

Federal Emergency Management Agency (FEMA) (n.d.). *Morgan Stanley case study.* Available at https://www.allrecipes.com/recipe/147363/the-real-mojito/print/?recipeType=Recipe& servings=1&isMetric=false (Last Accessed June 14, 2019).

Gantt, P., & Gantt, R. (2012). Disaster psychology: Dispelling the myths of panic. *Professional Safety, 11*(5), 42–49.

Goda, K., Kiyota, T., Pokhrel, R., Chiaro, G., Katagiri, T., Sharma, K., & Wilkinson, S. (2015). The 2015 Gorkha Nepal earthquake: Insights from earthquake survey damage. *Frontiers in Built Environment, 1*(8), 1–15.

Hamdani, K., Mills, C., Silva, J., & Battisto, J. (2018). *Puerto Rico small businesses and the 2017 hurricanes.* New York: Federal Reserve Bank of New York.

Henderson, M. G., Bent, R., Chen, Y., Delzanno, G. L., Jeffery, C. A., Jordanova, V. K., ... Woodroffe, J. R. (2017, December). Impacts of extreme space weather events on power grid infrastructure. In: *AGU fall meeting abstracts.*

Johnston, D. M., & Johnson, N. R. (1989). Role extension in disaster: Employee behavior at the Beverly Hills Supper Club fire. *Sociological Focus, 22*(1), 39–51.

Kearns, R., Stringer, L., Craig, J., Godette-Crawford, R., Black, P., Andra, D., & Winslow, J. (2017). Relying on the national mobile disaster hospital as a business continuity strategy in the aftermath of a tornado: The Louisville experience. *Journal of Business Continuity & Emergency Planning, 10*(3), 230–246.

Knobler, S., Mahmoud, A., Lemon, S. , et al. (2004). *Learning from SARS: Preparedness for the next disease outbreak: Workshop summary.* Washington, DC: National Academies.

Laboureur, D., Han, Z., Harding, B., Pineda, A., Pittman, W., Rosas, C., ... Mannan, M. S. (2016). Case study and lessons learned from the ammonium nitrate explosion at the West fertilizer facility. *Journal of Hazardous Materials, 308*, 164–172.

Langfield, T., Colthorpe, K., & Ainscough, L. (2017). Online instructional anatomy videos: Student usage, self-efficacy, and performance in upper limb regional anatomy assessment. *Anatomical Sciences Education, 11*(5), 461–470.

Marshall, M. I., Niehm, L. S., Sydnor, S. B., & Schrank, H. L. (2015). Predicting small business demise after a natural disaster: An analysis of pre-existing conditions. *Natural Hazards, 79* (1), 331–354.

Mathews, L. (2018). *Ransomware attack disrupts emergency services at Ohio hospital.* Forbes.com (Last Accessed May 19, 2019).

McDonald, G., Smith, N., Kim, J., Cronin, S., & Proctor, J. (2017). The spatial and temporal 'cost' of volcanic eruptions: Assessing economic impact, business inoperability, and spatial distribution of risk in the Auckland region, New Zealand. *Bulletin of Volcanology, 78*, 48.

Niall, P. A., Johnson, S., & Mueller, J. (2002). Updating the accounts: Global mortality of the 1918-1920 "Spanish" influenza pandemic. *Bulletin of the History of Medicine, 76*(1), 105–115.

Orchiston, C. (2013). Tourism business preparedness, resilience and disaster planning in a region of high seismic risk: The case of the Southern Alps, New Zealand. *Current Issues in Tourism, 16* (5), 477–494.

Paton, D., & Hill, R. (2006). Managing company risk and resilience through business continuity management. In D. Paton, & D. Johnston (Eds.), *Disaster resilience: An integrated approach* (pp. 249–266). Springfield, IL: Charles C. Thomas.

Perry, R., & Quarantelli, E. (2005). *What is a disaster? New answers to old questions.* Philadelphia: Xlibris.

Pescaroli, G., Turner, S., Gould, T., Alexander, D., & Wicks, R. (2017). *Cascading effects and escalations in wide-area power failures. A summary for emergency planners.* (UCL IRDR and London Resilience Special Report 2017-01)Institute for Risk and Disaster Reduction, University College London.

Phillips, B., Neal, D., Wikle, T., Subanthore, A., & Hyrapiet, S. (2008). Mass fatality management after the Indian Ocean Tsunami. *Disaster Prevention and Management, 17*(5), 681–697.

Phillips, B. D. (2015). *Disaster recovery.* Boca Raton: CRC Press.

Quarantelli, E. L. (1985). *Organizational behavior in disasters and implications for disaster planning.* Available at(1985). http://udspace.udel.edu/bitstream/handle/19716/136/PP247-Research%20Based%20Criteria.pdf?sequence=1. *(Last Accessed May 14, 2020).*

Quarantelli, E. L. (Ed.). (1998). *What is a disaster? Perspectives on the question.* London: Routledge.

Quarantelli, E. L., Lagadec, P., & Boin, A. (2006). A heuristic approach to future disasters and crises: New, old, and in-between types. In H. Rodriguez, E. L. Quarantelli, & R. Dynes (Eds.), *Handbook of disaster research* (pp. 16–41). New York: Springer.

Ramzy, A., & Bradsher, K. (2019). Explosion rocks industrial zone in eastern China, killing 47. *New York Times,*. March 21, 2019.

Trainor, J., & Barsky, L. (2011). *Reporting for duty? A synthesis of research on role conflict, strain, and abandonment among emergency responders during disasters and catastrophes.* (Miscellaneous Report 71)Newark, DE: Disaster Research Center.

Varandani, S. (2016). Cyclone Winston: Fiji's estimated cost of damages exceeds $470M, 10% of the island nation's total GDP. *International Business Times,*. Available at(2016). https://www.ibtimes.com/cyclone-winston-fijis-estimated-cost-damages-exceeds-470m-10-island-nations-total-gdp-2332151. (Last Accessed July 3, 2019).

Webb, G., Tierney, K., & Dahlhamer, J. (2000). Businesses and disasters: Empirical patterns and unanswered questions. *Natural Hazards Review, 1*(2), 83–90.

Webb, G., Tierney, K., & Dahlhamer, J. (2002). Predicting long-term business recovery from disaster: A comparison of the Loma Prieta earthquake and hurricane Andrew. *Global Environmental Change, Part B: Environmental Hazards, 4*, 45–58.

Yamazaki, F., Nishimura, A., & Ueno, Y. (1996). Estimation of human casualties due to urban earthquakes. *Prehospital Emergency Care, 8*(2), 217–222.

CHAPTER 2

Setting the stage

The challenges of business continuity planning

Business continuity planning (BCP) rarely falls onto anyone's desk unless it is the job for which they were hired. Usually, such hires occur in larger enterprises with funds sufficient to cover staffing costs. Most of the time, business owners, agency administrators, organizational leaders, and their employees undertake business continuity planning without such expertise, if they do take on such planning Yet, business continuity planning can be undertaken – even by novices – and reach successful outcomes. This volume keeps novices in mind as it lays out what planners need to know and how to accomplish specific tasks that result in a plan that can be executed in a disaster.

This chapter starts by unpacking and illustrating useful processes, approaches, and terms to orient everyone about the most important concerns. In the first section, readers will learn how to prioritize business continuity planning so that a successful effort can ensue. The second section explains key terms to help planners understand the consequences of an untoward event. The final section looks at how to put a planning team together and who should be on that team.

To start, we look first at the importance of getting people's attention and securing buy-in from top leadership. Barriers to getting that attention will be reviewed, with suggestions on how to convince people to plan for an event that might or might not happen – though the COVID-19 pandemic might have changed people's perspectives in 2020. The section then works through why the planning team is so important and what it will take to enable a high level of functioning within the team to complete the planning effort.

Getting people's attention

People do not like to think about negative events or terrible situations. Clearly, disasters and the loss of work, income, and opportunities fall into a negative framework – and people will avoid addressing the matter. People also tend to think "it won't happen here." After all, hurricanes turn away from anticipated

25

Business Continuity Planning. https://doi.org/10.1016/B978-0-12-813844-1.00007-5

landfalls at the last minute and tornados can be fickle where they strike – missing your business but taking the one down the street. In earthquake-prone areas, people worry about the "big one" while simultaneously thinking it to be statistically unlikely (Jones, 2019). Not everyone gets the flu, so why worry about a pandemic, although maybe that perspective has changed after COVID-19? Business owners and educators also think they lack resources to plan for a rainy day and prioritize other needs that occur more routinely (Dahlhamer & D'Souza, 1997; Webb, Tierney, & Dahlhamer, 2000, 2002).

Thus, the first critical action that any planner – or worried employee – needs to take, is to get people's attention. Often, the best time to motivate planning for a disaster is right after one has occurred. Ideally, planners will not be doing so after a direct hit on their business, although that does happen. But nearby or recent events can be used to alert business owners, managers, and leaders to think through "what if it happened here?."

One compelling way to get people's attention is to address how disasters affect the company's or university's financial bottom line (Hargis, Bird, & Phillips, 2014). As a routine, annual procedure, planners could look at insurance policies and examine the extent of coverage. What is the deductible? What is and is not covered? How much money has been set aside for deductibles? At some workplaces, deductibles can reach significant amounts into hundreds of thousands of dollars. How much can a company absorb financially – and what if the deductible needs to be met multiple times in a short period? What is not covered, and what would it take to replace those items or structures? Think also in terms of internal resources. For example, do computers and other technology (like an assembly operation or a mechanized warehouse distribution system) need to be addressed separately in an insurance policy? Does health care insurance sufficiently support employees who could be out for illness, injuries, or therapy? What about legal action from injured employees or customers? If employees need to take paid medical or personal leave, how will the company temporarily replace those employees?

Planners can further broaden the conversation by discussing how events impact businesses. For example, a fair amount of pandemic planning has been underway worldwide, but businesses generally lag in such planning. Pandemics can result in quarantines or closure, with government officials and health authorities holding the power to order such protective actions – but how ready are businesses, schools, and universities to comply and adapt? Imagine, for example, that an influenza outbreak affects the accounting and payroll section – how will payroll be completed? Losing key personnel in such a critical business function can compromise a business, even if only one section of the workplace is affected. Payroll units typically rely on data networks to deliver earnings. What happens if a cyberattack or malware compromise abilities to use the Internet or desktop software (Stanton, 2005)? Numerous businesses faced such a threat in 2018, when a global ransomware attack targeted banks worldwide. Health care industries have faced similar attacks when cybercriminals have held patient records hostage in exchange for ransom. Cybercrimes have also undermined government agencies, including electoral systems, that rely on computers.

Finally, what happens if you lose the workplace itself – but the employees are still able to work? How and where will the business relocate? How much time will it take until the business is operational and generating revenue again? What will that transition cost the business? Will insurance cover rebuilding, for example, in the case of a tornado or flood? Will the business have to adapt its size or operations to survive? These and other questions will be addressed in the second section of this chapter. For now, use the questions offered here to look around your business and consider the impacts. Ask co-workers what happened at the business during the COVID-19 pandemic or what they observed in other businesses. The information that you gather will lay a foundation to convince top management and fellow employees about the value of business continuity planning.

Securing a commitment from the CEO

What often grabs the attention of CEOs comes from how disasters disrupt the financial bottom line, which can be estimated by looking at potential structure and content losses as well as the expenses associated with being out of one's business site. After all, losing the site and contents that produce the business revenue stream potentially compromises everything – unless the business has a solid, effective business continuity plan. A classic emergency management strategy to gain the attention is to do a *loss estimation*, which will be covered further in Chapter 3. For this chapter, let's review a best practices example that occurred at the University of California-Berkeley, which rests on a significant earthquake fault line. Berkeley's loss estimation resulted in the CEO investing $1 billion (USD) to enable its enterprise to become more disaster resilient (please see Box 2.1 before continuing).

Can business continuity proceed with less cost than the billion dollars applied in the Berkeley example? Absolutely, and doing so could secure the CEO's attention. For example, free, downloadable software exists for business continuity planning with worksheets, videos, and guidance from the U.S.-based Federal Emergency Management Agency (see https://www.ready.gov/business-continuity-planning-suite, Last Accessed May 15, 2020, with similar templates easily found online elsewhere). Teams can walk through FEMA's stepwise materials on their own and call on area experts like emergency managers or university faculty to help. Chambers of Commerce, Safety Councils, and agencies that support area businesses can also be asked to sponsor and host planning events.

BOX 2.1 The University of Berkeley invests $1 billion.

The Hayward Fault lies underneath the University of California-Berkeley, one of the state's flagship universities and home to thousands of students, faculty, and staff. A significant adjacent community includes people who enjoy and attend events at the university and benefit from the economic benefits it generates regionally. Earthquakes do happen, and the area is generally considered overdue for a significant event. In the 1990s, Professor of Social Work Mary Comerio was tasked with assessing Berkeley's vulnerability to such shaking. She led efforts to assess the local risks, using a number of tools including a microzonation soil map, estimates of ground shaking, maps of campus infrastructure, building conditions and usage, and the number of occupants likely to be in them at peak time. Her team looked at repairs and downtime needed for occasional, rare, and very rare events, or three distinct scenarios. A very rare event, about a magnitude 7.5 earthquake, could result in a 7.5 magnitude earthquake costing $900 million in personal income losses and $861 million in sales to the three counties surrounding the campus. The team's detailed analysis uncovered that 75% of the university's research funds were concentrated in 17 buildings or about half of the campus space. Eleven of those buildings would be closed extensively in a 7.5 temblor, representing about 12% of the campus space. University research projects are usually externally funded, which brings a significant revenue stream into the institution to pay for faculty and student researchers, laboratory space, supplies, and utility costs as part of grant-generated indirect funding. Such external research funding can be significant.

The risk assessment and loss estimation numbers caught the eye of the university Chancellor, who invested $1 billion over a 20-year timeline to reduce potential impacts to the campus and the community including mitigation efforts and demolition. Since the 1990s, UC-Berkeley leadership has remained concerned about seismic risks. In 2019, the university chancellor announced a new seismic evaluation of 600 campus buildings using newer standards. Six buildings earned a "very poor rating" which will merit additional attention from the university. More will be detailed in 2020, so readers should check news.berkeley.edu for updates. Remember how Berkeley crafted these three scenarios, which will become useful in Chapter 7.

Sources: Comerio, M. C. (2000). The economic benefits of a disaster-resistant university. Berkeley, CA: University of California Berkeley; Comerio, M. (2006). Estimating downtime in loss modeling. Earthquake Spectra, 22(2), 349–365; UC Berkeley Public Affairs (2019). What you should know about seismic issues on Berkeley's campus. Available at https://news.berkeley.edu/2019/08/28/what-you-should-know-about-seismic-issues-on-berkeleys-campus/ (Last Viewed October 14, 2019).

In short, businesses of all kinds can participate in and make choices about BCP efforts, with an investment ranging from minimal cost to a considerable amount. All business enterprises will have to decide what kind of risks they can tolerate versus what they can afford to do. Yet, even a little planning can make a difference, like creating an inexpensive "go-kit" to use when disaster threatens. That was what the Leidenheimer Bakery did in New Orleans (FEMA, 2011). The owner faced hurricane Katrina in 2005, an event that flooded the city and damaged not only businesses but employee and customer homes, utilities, and transportation arteries. The bakery shut down on a Sunday, with Katrina bearing down the same night. While evacuating, the owner notified insurance, accounting, and legal providers and alerted customers. He could do this because the bakery's evacuation go-kit included financial and payroll records, utility contacts, and phone lists for customers and employees. Next, the bakery regrouped in a temporary location during their displacement. The owner forwarded phones to a satellite office in a nearby city while waiting for New Orleans to be dewatered. Given that Katrina flooded a city of over half a million people, Katrina caused direct damage to the

150-year old bakery. The owner organized carpools for displaced employees, temporarily produced its bread out of a Chicago site to reduce downtime and re-opened after a massive clean-up. They were ready and reopened successfully, even though they were not professional emergency managers, and remain in business over 15 years later.

Lack of knowledge or understanding

Probably the next challenge for planners to overcome is to understand how different disasters could impact businesses. A useful first task, then, will be a hazard identification and risk assessment, which will be explained further in Chapter 3. Local emergency management agencies will know how to do a hazard identification, which is a review of historic and emerging threats, their local frequency, and potential impacts. It would be wise to invite an emergency manager to present that information to your team as a step-saving measure – but businesses also must do their own homework. In the University of California Berkeley example mentioned earlier, planners realized that site specific earthquakes would impact laboratories and introduced highly localized mitigation (meaning risk reduction) measures for laboratory shelves holding glassware and chemicals. Readers will work through a hazard identification process in more detail in Chapter 3. In doing that work, involving your team is key because employees are familiar with workplace procedures, policies, and resources. They may already have created and used workarounds in smaller scale events and know how to address threats to productivity.

Risk assessment (see Chapter 3) involves another body of knowledge that weighs the probabilities and consequences of a disaster. Such an assessment was part of the ingenious effort at Berkeley, which created low, medium, and high probability earthquake events and then contrasted those scenarios with loss estimations. Their risk assessment effort resulted in the University determining where and how it could best afford to reduce losses for the university. They invested in mitigation, understanding that the payoff might occur well beyond their own tenure as university employees. It was not an easy choice, but they made the best determination they could on a thorough examination of relevant facts followed by careful consideration.

Part of understanding risk requires weighing various outcomes and potential impacts. A decision will have to be made about what level of a disaster to plan for, understanding that every workplace must make judgement calls vis-à-vis their resources. We all (businesses, governments, universities, organizations, and agencies) will have to make hard choices about how much to invest in protective measures because we all have limits to our resources. We reach those choices through motivating stakeholders to participate in the business continuity planning process.

Educating and involving people on planning as a process

Anyone tasked with business continuity planning will need to help people understand why they need each other on the planning team. Many businesses rely on key personnel to shoulder critical tasks, from hiring and paying people to supervising and carrying out specific tasks. Knowledge lies within the people who take on those roles and their experience matters. A planning team must therefore engage actively with a broad set of people who can offer valuable insights from their own work experience. We all know that policies drive much of what we do and how we do it but making that policy work as it should requires knowledge specific to the task. People hold that knowledge and involving them in walking through the planning process helps to capture those valuable lessons learned in doing the job. As an example, consider the needs of employees after a disaster – human resources holds knowledge about policies, procedures and leave to help employees who are injured or need time to be a caregiver and would benefit from employee assistance programs. Involving the human resources team would be essential to completing a robust BCP (see also Chapter 6). Other areas in a workplace also hold important knowledge, so casting a wide net of involvement improves any planning effort. Further, disasters tend to disrupt routine policies and procedures, requiring sometimes innovative approaches and adaptation in how things are done (Kendra & Wachtendorf, 2007; Neal & Phillips, 1995; Srivastava & Shaw, 2015). By involving a wide array of people thinking about a challenge, planners can surface more solutions.

Employees in various units or offices may see risks differently than a colleague because of the responsibilities they carry or the point of view that they bring. Lead planners need to bring that expertise into play, but not everyone feels comfortable speaking up in a group. Social interaction, such as you would see in a planning team, can influence how people behave. Participants sometimes assume that others think or feel similarly and that their opinion is inconsistent with the group. Social psychologists call this groupthink, where people do not speak up because they assume unanimity. In such instances, people self-mute. In preparing for a disaster, not speaking up can be dangerous or even deadly which is what researchers found in discussions surrounding the failed Bay of Pigs invasion in 1962 and the Challenger explosion in 1987 (Hughes & White, 2010; Janis, 1971). A truly participatory process, where people feel free to offer their points of view without automatic rejection must be built carefully from the first meeting and nurtured throughout the planning process.

Several strategies can be undertaken to increase participation:

- Set the initial expectation at a first meeting that everyone needs to participate.
- Encourage people who do not speak up to do so.

- Require respect and active listening for and from everyone present.
- Go around the table and ask everyone to offer a different point of view to surface divergent perspectives.
- Use alternative means to gather information, such as collecting opinions on anonymous pieces of paper. Post or share them so that planners can review them and spot differences and new ideas.
- Break the planning team into smaller groups to host focused discussions and have the smaller groups report back.
- Monitor people who like to talk a lot and moderate the discussion so that others can join in.
- Observe carefully where hierarchy influences discussion, such as the presence of an executive who may unintentionally stifle input from a custodian or people who fear to speak up in front of their boss.
- Reward people for their participation with verbal praise and through thoughtful emails to their supervisors. At the end of the planning process, publicly thank all participants for their work and provide them with appropriate recognition, such as a formal letter in their employee record, a certificate, or a plaque.

Getting people to participate is only part of the process, which will require everyone to join in the various steps and stages of BCP. The team needs to commit to going through the process so that every part is reviewed by people across the business for holes, problems, issues, and – perhaps most importantly – ways to adapt when disaster strikes. People hold many of the insights and answers inside of them, and it is the job of the planning team leader to bring that knowledge to the planning table. Harnessing that knowledge and directing it into useful efforts requires understanding some basic terms to focus the team, which serves as the purpose of the next section.

The effects of disasters on businesses

In this section, readers will learn more about the key concepts that drive business continuity planning: direct and indirect impacts, downtime, and displacement as well as how disasters impact employees, customers, and business survival. *Direct impacts* occur when a disaster hits a business and surrounding community head-on, such as the 2004 tsunami wave that crushed the main commercial sector of Vailankanni, India or the 1984 gas leak that killed over 10,000 employees and residents of Bhopal, India. *Indirect impacts* also occur, such as when climate change results in droughts that undermine agricultural productivity or cause wildfires that destroy entire towns. Both direct and indirect impacts may cause *downtime*, which means that a business cannot operate or faces a slowdown (Chang & Falit-Baiamonte, 2002). The operating room may cease elective surgeries, or the bank might have to close because both rely on power. *Displacement* occurs when businesses relocate either temporarily or permanently because of the disaster impacts. Ideally, a workplace will want to reduce such impacts because they want to retain employees, serve their customers and clients, and keep the business operating.

Direct impacts

Direct impacts catch our attention because of the visual damage we can see. Tornadoes take out power lines, hurricanes flood roads, and earthquakes compromise buildings (see Fig. 2.1). It is easy to see and understand the direct impacts, which can be used to focus people's attention on why continuity planning matters. When a disaster strikes businesses elsewhere, planners can use the moment to educate people on similar risks in their locations. Consider, for example, these kinds of impacts on businesses:

- 1989, a coronal mass ejection (solar storm) caused a 12-h blackout across the province of Québec in Canada. Effects included employees being unable to go to work without abilities to use mass transit, elevators, or computers. Later in 1989, the Toronto Stock Exchange stopped for a while

FIGURE 2.1
Dust from the World Trade Center attack *Andrea Booher/FEMA News Photo*

during a similar event, which disrupted trading. Although effects also reached the U.S., the power grid held this time. Worry continues that similar storms could disable the U.S. eastern seaboard power grid in a future event.

- 1999, an ice storm crippled a portion of Canada and reached into the northern U.S. The heavy ice crushed utility poles and disabled transmission towers, resulting in business closures that lasted for weeks. Canadian military forces had to be deployed as people went without power and heat. Bridges, tunnels, and roads had to be closed and utilities like pumping stations stopped functioning.
- 2001, the terror attack of September 11th claimed the lives of over 3000 employees and first responders. Three sites sustained damage, particularly in New York City where multi-story buildings collapsed, sending dust and debris into surrounding businesses, homes, and schools. The New York Stock Exchange, major sporting events, and many businesses stopped. In New York, the hospitality industry was particularly affected and airline industries suffered significant economic losses.
- 2004, the Indian Ocean tsunami destroyed an entire commercial sector in Vailankanni, India and undermined area fisheries for years. In Thailand, thousands of tourists died when debris-filled waves pushed inland also damaging the fishing, agricultural, and hospitality industries. The tsunami damaged business areas in a dozen nations and claimed over 300,000 lives.
- 2011, an EF5 tornado ravaged Joplin, Missouri, heavily damaging Main Street businesses as well as industry-specific sites like nursing homes, schools, restaurants, and the local hospital which had to be evacuated. Health care employees scrambled to use available vehicles, including their personal cars, to transport critically ill and injured patients to a safer location. Simultaneously, hundreds of injured people in the area produced a patient surge the hospital had to meet.
- 2013, Typhoon Haiyan tore through the Philippines, leaving 6 million workers with lost livelihoods from affected farms and small businesses. With crops damaged and roads impassable, employers and employees struggled to restore income and re-open businesses, in an event that cost the nation billions of dollars.
- 2018, the Camp Fire near Paradise, California incinerated over 500 businesses and nearly 14,000 homes. Additional structures including government offices, area farms and homesteads, and schools succumbed to the wildfire as well. Health care industries had to set up separate command centers and reorganize shifts to cover patient care.
- 2018, the Kilauea volcano in Hawaii erupted again, undermining tourism, and resulting in cancelled vacations that caused significant losses to hotels, restaurants, airlines, and related industries.

- 2019, hurricane Dorian hovered over isolated islands in the Bahamas, destroying business and government areas, residential neighborhoods, roads, and ports. About a month later, estimated economic impacts indicated that half of the nation's gross national product was directly affected (International Medical Corps, 2019).
- 2019, the city of Atlanta faced a massive cyberattack that stopped the use of computers across city operations, affected Internet connectivity at the international airport, and cost the city millions of dollars.

Direct impacts threaten a business so much that it cannot function well or at all. When flood waters enter a building or high wind throws projectiles through windows, business stops. Damage to the building or site, or the systems, production lines, restaurant tables, emergency rooms, conference centers, or computers means that people cannot work, customers or clients cannot be supported, people will not be paid, contracts and orders will go unmet, and businesses may fail. A direct impact on a business, school, or agency may also affect business sectors that rely on the main industries. When those key businesses stop, collateral damage can occur when employers and employees stop producing taxes and can't buy gas or pay for gas, food, childcare, rent, mortgages, and loans.

Direct impacts provide compelling images and evidence that a business needs to prepare for the consequences of a direct hit, to reduce what effects they can, and to be ready to pick up the pieces and continue. By using evidence of impacts on similar businesses, a planning team can bring attention to the value of business continuity planning. So, planners should spend time researching how disasters have affected businesses in their area but also from the same kinds of hazards they have in their area. Even though a flood has not affected businesses in the area, it is possible that it could – given that it is the most common disaster worldwide – and it could happen tomorrow even though it has not done so in 500 years. Conducting the hazard identification, risk assessment, and loss estimation for direct impacts (see Chapter 3) will help build a compelling case to get the CEO's attention and focus a planning team on direct impacts.

Indirect impacts

Perhaps one of the reasons that people fail to address business continuity planning is because they focus on the direct impacts and think "it won't happen here." We often think about the damage that a tornado or earthquake can cause or worry about a terror attack at the workplace. These kinds of direct impacts certainly require attention – but so do indirect impacts, which may be harder to spot (Graveline & Grémont, 2017). The tornado could miss your business but take down utility sources that power computers, lights, operating rooms, ovens, assembly lines, gas stations, Internet service, and banks. An earthquake can damage transportation arteries used to deliver goods and services, or fracture underground infrastructure necessary for sewers or subways. Emergency managers call such consequences indirect impacts.

Indirect impacts can be as significant as direct impacts because of lost opportunities to generate revenue, serve customers, care for patients, create products, and handle even routine matters. One classic study done on the 1994 Northridge earthquake (Dahlhamer & Tierney, 1998) found that 60% of area businesses lost electricity, 50% lost phone service, 20% lost water and 17% lost natural gas. The consequences? Over 40% reported reduced customer traffic and 25% had trouble receiving deliveries. Because of the earthquake, 60% said employees could not get to work. Half of the owners and managers surveyed also had personal property damage making it difficult to choose between work and home. In 1993, a Midwestern flood along the Mississippi River affected nine states – and it closed the Mississippi River, a major cargo transportation route (Quarantelli, 1985). In Des Moines, Iowa, businesses surveyed spent an average of 12 days without electricity, phones, sewer and wastewater, and water (Dahlhamer & Tierney, 1998). A 2010 flood that severely impacted nine districts in Pakistan affected over 300,000 small businesses with loss of utilities, lack of water, and disrupted transportation routes (Asgary, Anjum, & Azimi, 2012). Nearly one-third had lifeline disruptions and almost 1 in 5 suffered disruptions in their supply chain. The 2011 earthquake and tsunami in Japan disrupted both domestic and global supply chains that caused a daunting recovery process (Leelawat, Suppasri, & Imamura, 2015; Tokui, Kawasaki, & Miyagawa, 2017). Globally, prices for flash memory – used in LCD parts and materials – increased by 20% (Park, Hong, & Roh, 2013). Damage to one Japanese company that supplied pigments for car paint caused automotive plant shutdowns for major car and trucking manufacturers in the U.S. How long could your business survive indirect impacts?

Regardless of location, small businesses seem to fare worse with both direct and indirect impacts, especially depending on their resources (Torres, Marshall, & Sydnor, 2019). A study of how hurricane Katrina impacted 499 businesses in Mississippi revealed that small businesses varied in how they had prepared for both direct and indirect impacts (Josephson, Schrank, & Marshall, 2017). Small business owners who rented their locations had to rely on the building owners to mitigate structural impacts and faced displacement as a result. Conversely, long-time owners had reduced mitigation activities like buying insurance, possibly because they had paid off mortgages that required insurance. However, some factors influenced readiness. Living in closer proximity to the coast, where storm surge and wind damage would be highest, was associated with better preparedness as a likely outcome of regularly experiencing coastal storms. Interestingly, female owners were more likely to prepare for disaster with an emphasis on structural mitigation. Business owners seem to worry the most about a direct impact. Yet the indirect impacts from loss of utilities or transportation arteries can also influence a business as can the effects of stay at home orders during a pandemic. We call the outcome of these direct and indirect impacts "downtime."

Downtime

How long can a business be closed without serious economic impacts? Both direct and indirect impacts can produce downtime which is the number of hours, days, or months that a business is unable to fulfil its mission, deliver products, support the local economy, and pay workers. Imagine a tornado that tears part of the roof off a local childcare site – the workers will likely lose their jobs for several months while the damage is repaired. Meanwhile, families who rely on childcare will need a new location, in a search that may disrupt their own workdays and impact their personal budgets. The owner of the childcare business may face loss of their clientele until they can re-open, and then need to spend marketing dollars to secure new clients.

Most people have experienced an IT disruption which causes downtime to take place. Email may slow or become unavailable. Internet connections can be lost as well, sometimes for significant amounts of time. With so many businesses dependent on such technology, the consequences for lost time, lost revenue, and impacts to services and clients can be significant. It is probably best to assume that downtime of some kind will occur (Coffey, Postal, Houston, & McKeeby, 2016).

As an example, consider the health care industry. Hospitals rely on Internet connections and software to capture and share patient information including critical information that can be lifesaving. The National Institutes of Health Clinical Center (a clinical research hospital) experienced a 33-h downtime on May 13, 2010 due to a hardware failure (Coffey et al., 2016). The problem also affected their primary and backup databases with all clinical information lost to the 3200 users. Resolving the situation required waiting for a replacement part, then restoring a trusted backup. Although the Center had a downtime plan, they used the event to rethink what to do. Consequently, they formed the Downtime Communication, Education, and Training Committee. One new solution they put into place was a downtime toolkit with organized file folders tied to critical paperwork that had to be collected. Each patient area received a toolkit with appropriate education followed by quarterly downtime drills (Coffey et al., 2016).

In 2017, a study of 59 large health care systems in the U.S. revealed that 96% had experienced downtime in a 1-year period. However, of those 59 systems, only 28% had exercised a downtime plan (Kashiwagi et al., 2017). Using the acronym CLEAR, which resonates within the medical community as a term related to cardiac defibrillator use, the study's authors crafted a work area specific downtime effort. C, check and communicate the problem, tasked staff with knowing how to check IT systems and report outages. L, for locating the downtime plan, involved staff in using the right plans and forms to initiate a response. E, which asked staff to establish a different way to meet patient processes (admissions, therapy, etc.), resulted in continuity of care. A, meant that staff knew how to authorize and activate a plan while R meant that staff could recover by entering data into the system after the downtime ended. This CLEAR procedure restored routines for patient care teams so they could continue with operations as close to normal as possible. The organization also stood behind the effort, which resulted in widespread education, preparation, and training.

Displacement

Another concern that planners must address is to consider the costs and needs of being displaced. If a business is unable to function in its present location, what kind of new location – temporary or permanent – will be needed? How long will it take to move there, which needs to be considered, as well as the length of time to get back up and running? How will customers find the new location? Will employees be able to reach the new site, or will the business lose valued workers? What will the site be like – and what can be accommodated there?

One study contrasted the 1989 Loma Prieta (California) earthquake with the 1992 impact of hurricane Andrew in Florida. Close to 1000 businesses responded to surveys in each disaster (Wasileski, Rodríguez, & Diaz, 2011). Fewer businesses relocated for the Loma Prieta earthquake, although the downtown sector that was heavily damaged in Santa Cruz, California received extensive local support to re-open via temporary sites. Some factors that did influence relocation in both disasters included renting a space. If a business had to rely on a building owner, they faced a higher chance of having to relocate. For Loma Prieta, being in a building with unreinforced masonry increased relocation changes when the earthquake damaged the structure. Being in a manufacturing or retail sector also increased relocation likelihood. For Andrew, disruption from loss of utilities or damage to building contents increased relocation possibilities. Business owners who also faced economic challenges before the hurricane damaged the area saw increased relocation chances, most likely because of limited resources (Wasileski et al., 2011).

Overall, renters may face more challenges with displacement which can worsen with the type of hazard and the type of building the business is in. The sector of the business, such as retail, coupled with its location can worsen matters. Pre-existing conditions also influence what happens including the general economy and the business's assets (Marshall, Niehm, Sydnor, & Schrank, 2015), all of which will result in varying impacts on employees who want the business to survive and thrive.

Impacts on employees

Workplaces invest a lot of time, money, and effort in finding, training, and retaining valued employees. How might a disaster affect one's employees? The first thing that usually comes to people's mind is payroll – how long can a business pay their valued employees without generating revenue from their labor? The answer lies in how well a business has assessed risk and amassed resources for a rainy day including assets that can be liquidated and cash on hand as well as the business continuity plan that lays out how to adapt when disaster strikes.

Disasters can also impact employees in other ways besides income. They could become injured in the event or even lose their lives. Companies will need to be sure that health care coverage sufficiently insures employees and also consider ways to support employees who may be struggling to pay deductibles, bills denied by the insurance provider, co-pays, and other costs not covered by insurance like child care, wages lost to a caregiver, or home health support. Long-term rehabilitation may be needed, and companies might want to find ways to support those costs for employees. Companies may also need to document what happened should legal action occur either because of their own negligence or if a criminal act like terrorism caused the disaster.

The good news is that employees step up, over and over, in a disaster. A business continuity plan can anticipate this level of support, integrate it into their plan, and develop a means to recognize and reward employees from those who would receive hazard pay to those who would merit company-wide recognition through other benefits.

Impacts on customers

Several kinds of impacts can be anticipated on customers, clients, patients, and students. The most important one concerns customer safety while in a building or working with a business. Companies hold legal and ethical responsibilities to safeguard employees and others inside their buildings, including from natural disasters and active aggressors. In 2019, the MGM Grand settled a lawsuit on the mass shooting in Las Vegas where 58 people died, and hundreds sustained injuries. The settlement allowed for $800 million to victim claims that the hotel had allowed the killer to stockpile weapons in his hotel room. Lawsuits from COVID-19 will also reveal the kinds of impacts on customers, particularly residents of nursing homes where the virus claimed a disproportionate number of lives. Lawsuits have also ensued from communities where people of color had been denied a test or died from not being prioritized for hospitalization.

Also consider the ways in which customers may be affected by crimes that undermine businesses using private data. Cybercrimes clearly affect customers and can cost businesses a significant amount of time and money, from providing identify theft protection to legal action against the company. If people feel their information is not safe with a company, they will not patronize that site. Customers may also steer away from businesses they believe have not responded to a disaster, including cybercrimes, from an accountable and ethical business practice. Accusations about poor employee care of essential workers erupted during COVID-19, leading to large protests and strikes against some businesses believed to have failed in protecting their workers. The same problem emerges for non-profit agencies when donors believe their funds have not been used for the intended purpose or they hear of an employee who mishandled a financial matter.

Customers may also find that disrupted traffic patterns deter them from a preferred business and lead them to new vendors and providers. Transportation impacts, visible signs of damage, or an assumption that a business has closed may reduce a customer base. Depending on the disaster, fear over the safety of a product or location (like food during an epidemic or a hazardous materials spill near a business sector) may reduce customer support. Customers need to feel safe, supported, and able to access a product or service to be retained. Businesses may have to launch proactive campaigns to keep or lure back their customer base.

Businesses that fail

What business continuity planning comes down to is protecting the financial bottom line to stay afloat. Businesses that fail have typically failed to prepare, including planning, and in setting aside resources for recovery. In a study of CPA firms affected by hurricane Sandy on Staten Island, New York, most did not have a plan before the storm or know what a business continuity plan looked like (Scarinci, 2014). Those who had some plans in place for a disaster had failed to test their plans and discovered they lacked access to back-up sites during displacement and had left key information and resources in their offices. Cell phones also failed due to extended power outages, and many firms had key phone contacts on those cells (Scarinci, 2014). In some studies (Marshall et al., 2015), disaster experience motivates business continuity planning and disaster preparedness. Yet even in Thailand, where the 2004 tsunami devastated coastal businesses from tourism to fishing, business continuity planning still lags, which may be related to available funds and time. In a study of 136 small and medium Thai businesses, nearly half (44%) had experienced a disaster in the 10 years following the tsunami. Yet, most had not completed any disaster preparedness measures such as a plan or employee training (Kato & Charoenrat, 2018). Still, a promising number did have a written business continuity plan and another 75.5% said that they were developing one.

Business size and economic resources appear to be the key factor explaining such findings in Thailand, which is consistent in other nations as well

(Karim, 2011; Kato & Charoenrat, 2018). In short, larger businesses have more resources that enable them to rebound from a disaster. That does not mean that a small business cannot succeed, because many can and do survive after a disaster. To be more resilient, start with, finish, and train on a plan suitable to your situation, location, and resources.

Essential actions

The business continuity planning process begins with a few key steps:

- *Get a team together.* Cast a wide net when establishing a planning team. Choose someone who can facilitate discussion and groups well, because their leadership will make all the difference in producing a good result. Involve people broadly across the enterprise and think through the horizontal and vertical organization of a business. Planning should include people from diverse points of view within the system because each will be able to offer perspectives from where they are located – indeed, ask them to lead from where they are located. Include people new to the business (because they ask great questions being new) and people who are veterans of the enterprise (because they know how it used to be done, which could be a great source of information). Diversity matters as well in bringing people on to the team who represent people with disabilities, people of color, and from various genders. They will bring valuable insights to the planning table from where they work.
- *Establish a timeline.* Creating a plan can vary from a short and focused effort sufficient for many small businesses to a more extensive effort that will require multiple kinds of plans across a massive system. Scale matters in establishing a timeline, and business continuity planning needs to be realistic. Teams will need to accommodate for people with heavy workloads who cannot complete a task under a quick deadline as well as for people who may be out sick, on business travel, or on vacation. Many things interrupt timelines, so look at the tasks in this volume and set out reasonable times to complete those steps. It may be valuable to look down the road at the start of a specific hazard season, such as March 1 when tornado awareness season begins in the U.S. and have that as a goal to complete the plans.
- *Kick off the BCP effort.* Pull people together to talk about the planning process and to help them get to know each other. Involve the CEO, who should talk about the importance of their effort for the entire enterprise and set a standard of expectation that everyone needs to participate fully and equally. Talk through the timeline to identify times that may enable the team to fulfill a specific task or could slow them down (e.g., annual holidays). Invite an inspirational speaker such as a local emergency

manager or someone from a similar business who experienced a disaster. Bring them to the team in person or virtually. Individually recognize each person present and their expertise and explain why each offers value to the planning team.

- *Build and maintain momentum.* Business continuity planning is rarely at the top of anyone's to-do list, so a planning team coordinator or leader will need to keep people on task at a reasonable pace. After all, people are probably doing this on top of their other duties and will need to be mentored, supported, and encouraged. Many team members will also be learning about business continuity planning which can feel overwhelming and unfamiliar at first. Celebrate accomplishments by acknowledging that the team has hit a goal in the timeline or completed a specific task like a hazard identification. Share that information with others in the business so that they know what is happening and can support and recognize the planning team. Keep going because, when a disaster hits like the world experienced with COVID-19, you will want the plan ready to put into operation.

References

Asgary, A., Anjum, M., & Azimi, N. (2012). Disaster recovery and business continuity after the 2010 flood in Pakistan: Case of small businesses. *International Journal of Disaster Risk Reduction, 2,* 46–56.

Chang, S. E., & Falit-Baiamonte, A. (2002). Disaster vulnerability of businesses in the 2001 Nisqually earthquake. *Environmental Hazards, 4,* 59–71.

Coffey, P., Postal, S., Houston, S., & McKeeby, J. (2016). Lessons learned from an electronic health record downtime. *Perspectives in Health Information Management,*(Summer). Available at (2016). https://perspectives.ahima.org/lessons-learned-from-an-electronic-health-record-downtime/. (Last Accessed November 25, 2019).

Dahlhamer, J. M., & D'Souza, M. J. (1997). Determinants of business disaster preparedness. *International Journal of Mass Emergencies and Disasters, 15*(2), 265–283.

Dahlhamer, J. M., & Tierney, K. J. (1998). Rebounding from disruptive events: Business recovery following the Northridge earthquake. *Sociological Spectrum, 18*(2), 121–141.

FEMA (2011). *Sandy Whann bakes in a large helping of preparation. Available at (2011). https://www. fema.gov/el/media-library/assets/documents/89548. (Last Accessed October 14, 2019).*

Graveline, N., & Grémont, M. (2017). Measuring and understanding the microeconomic resilience of businesses to lifeline service interruptions due to natural disasters. *International Journal of Disaster Risk Reduction, 24,* 526–538.

Hargis, B., Bird, L., & Phillips, B. (2014). Building resilience to natural disasters across the campus ecosystem. In *Managing the unthinkable: Crisis preparation and response for campus leaders* (pp. 18–36). Washington, DC: American Council on Education.

Hughes, P., & White, E. (2010). The Space Shuttle Challenger disaster: A classic example of group-think. *Ethics & Critical Thinking Journal, 2010*(3), 63.

International Medical Corps (2019). *Hurricane Dorian situation report #16. Available at(2019). https://reliefweb.int/report/bahamas/hurricane-dorian-situation-report-16-november-19-2019. (Last Accessed November 21, 2019).*

Janis, I. L. (1971). Groupthink. *Psychology Today, 5*(6), 43–46.

Jones, L. (2019). *The big ones: How natural disasters have shaped us (and what we can do about them)*. NY: Doubleday.

Josephson, A., Schrank, H., & Marshall, M. (2017). Assessing preparedness of small businesses for hurricane disasters: Analysis of pre-disaster owner, business and location characteristics. *International Journal of Disaster Risk Reduction, 23*, 25–35.

Karim, A. (2011). Business disaster preparedness: An empirical study for measuring the factors of business continuity to face disaster. *International Journal of Business and Social Science, 2*(18), 183–192.

Kashiwagi, D., Sexton, M., Graves, C. , et al. (2017). All CLEAR? Preparing for IT downtime. *American Journal of Medical Quality, 32*(5), 547–551.

Kato, M., & Charoenrat, T. (2018). Business continuity management of small and medium sized enterprises: Evidence from Thailand. *International Journal of Disaster Risk Reduction, 27*, 577–587.

Kendra, J., & Wachtendorf, T. (2007). Improvisation, creativity, and the art of emergency management. *Understanding and responding to terrorism* (pp. 324–335)Vol. 19, (pp. 324–335). .

Leelawat, N., Suppasri, A., & Imamura, F. (2015). Disaster recovery and reconstruction following the 2011 Great East Japan earthquake and tsunami: A business process management perspective. *International Journal of Disaster Risk Science, 6*, 310–314.

Marshall, M. I., Niehm, L. S., Sydnor, S. B., & Schrank, H. L. (2015). Predicting small business demise after a natural disaster: an analysis of pre-existing conditions. *Natural Hazards, 79*(1), 331–354.

Neal, D. M., & Phillips, B. D. (1995). Effective emergency management: Reconsidering the bureaucratic approach. *Disasters, 19*(4), 327–337.

Park, Y., Hong, P., & Roh, J. (2013). Supply chain lessons from the catastrophic natural disaster in Japan. *Business Horizons, 56*, 75–85.

Quarantelli, E. L. (1985). What is disaster? The need for clarification in definition and conceptualization in research. *Disasters and Mental Health: Selected, 10*, 41–73.

Scarinci, C. (2014). Contingency planning and disaster recovery after hurricane Sandy. *The CPA Journal, 84*(6), 60–63.

Srivastava, N., & Shaw, R. (2015). Occupational resilience to floods across the urban-rural domain in Greater Ahmedabad, India. *International Journal of Disaster Risk Reduction, 12*, 81–92.

Stanton, R. (2005). Beyond disaster recovery: The benefits of business continuity. *Computer Fraud & Security*, 18–19.

Tokui, J., Kawasaki, K., & Miyagawa, T. (2017). The economic impact of supply chain disruptions from the Great East-Japan earthquake. *Japan and the World Economy, 41*, 59–70.

Torres, A., Marshall, M., & Sydnor, S. (2019). Does social capital pay off? The case of small business resilience after hurricane Katrina. *Journal of Contingencies & Crisis Management, 27*, 168–181.

Wasileski, G., Rodríguez, H., & Diaz, W. (2011). Business closure and relocation: A comparative analysis of the Loma Prieta earthquake and hurricane Andrew. *Disasters, 35*(1), 102–129.

Webb, G., Tierney, K., & Dahlhamer, J. (2000). Businesses and disasters: Empirical patterns and unanswered questions. *Natural Hazards Review, 1*(2), 83–90.

Webb, G., Tierney, K., & Dahlhamer, J. (2002). Predicting long-term business recovery from disaster: A comparison of the Loma Prieta earthquake and hurricane Andrew. *Environmental Hazards, 4*, 45–58.

CHAPTER 3

Pre-planning steps to launch BCP

Introduction

In this chapter, planning teams will gather information as essential pre-planning actions needed to write a business continuity plan. Pre-planning steps for business continuity include identifying what kinds of hazards a business may face, how to assess those risks, and ways to determine potential losses. Once these activities have been completed, businesses will know what kinds of impacts to expect, whether from a flood, a cyber attack, or pandemic. Loss estimation in particular carries the ability to get the attention of the head of the business, because those losses will likely determine what steps can and must be taken to increase business resilience, to support employees, and to recover from minor to major blows to revenue, production, and viability. While each of these procedures may seem daunting, even small businesses or agencies that lack expertise or an emergency management staff can walk through the steps outlined in this chapter (Josephson, Schrank, & Marshall, 2017; Kato & Charoenrat, 2018). We will start with discerning what kind of disaster might happen, why the kind of disaster matters, and how to consider and weigh that information.

Hazard identification

The purpose of a hazard identification is to reveal potential risks in a geographic area (Chartres, Bero, & Norris, 2019). The good news is that many jurisdictions where you work will have already conducted a hazard identification. By calling the local emergency management agency, businesses and agencies may be able to secure at least a partial set of known hazards such as earthquake zones, seasonal events like hurricanes, or even the risk of terrorism. Emergency managers may be very willing to share this information and may have posted it on their websites as many of these assessments are part of plans required federal funding opportunities in the U.S. Other websites may prove useful as well in identifying local or regional hazards that could imperil a business or agency (see Box 3.1).

Business Continuity Planning. https://doi.org/10.1016/B978-0-12-813844-1.00004-X

BOX 3.1 Sources of hazard information (all sources Last Accessed May 1, 2020)

- For a trustworthy, international database on disasters, visit https://www.emdat.be/. EM-DAT was created by the **Centre for Research on the Epidemiology of Disasters (CRED), the World Health Organization (WHO), and the Belgian Government.**
- For information on emerging and historic pandemics, visit the World Health Organization at https://www.who.int/.
- General governmental information sources may provide useful historic and current information such as Public Health Canada, with disaster information available at https://www.publicsafety.gc.ca/cnt/rsrcs/cndn-dsstr-dtbs/index-en.aspx.
- Hazard-specific information may be useful as well. For example, the National Weather Service in the U.S., available at https://www.weather.gov/ has daily and historic reports as does the National Hurricane Center at https://www.nhc.noaa.gov/. Just search for information like "earthquake risks" and you will hit on reliable sources like https://earthquake.usgs.gov/earthquakes/map/ or https://www.jma.go.jp/en/quake/.
- Regional hazard information can also be easily looked at by accessing websites like those linked from here: https://www.nhc.noaa.gov/aboutrsmc.shtml.
- International sources can also be useful like the International Tsunami Information Center at http://itic.ioc-unesco.org/index.php.

Local information sources can also be very helpful in discerning local hazards and the risks they pose. Consult with your local, state, or regional:

- Emergency Management Agency, which may have already completed a hazard identification and risk assessment for your area. They may have posted it on their website or can make it available, such as the one found at https://www.in.gov/dhs/mitigation.htm (Last Accessed May 12, 2020). The state, provincial, or regional office tasked with emergency or disaster management has probably conducted hazard analyses as well. They may have posted that information on websites or can make it available for business continuity planners.
- Public Health Agency, which covers infectious diseases and risks and works in concert with other first responders and emergency managers when events warrant.
- Public Works and or Planning Department, which will know where flood risks exist and will be able to access maps for various hazards. They may have a GIS system that can provide additional information for your consideration.
- Library, and ask the reference librarian for their files and resources on area storms and disasters.
- Historical societies, which often contain useful documents, photos, and archives on prior disasters that might bring a risk to life visually and with detail.
- Area universities may offer courses, certificates, or training on natural hazards, terrorism, or hazardous materials. Call and ask who has expertise that might be offered as part of the university's extension service or community engagement efforts.
- Several official government websites offer information useful to a specific region. By checking with the official weather service or with an agency tasked with something as specific as earthquakes, hurricanes, cyclones, or tornadoes, historic information can be secured potentially including probabilities of repeat events.

Those risks may be very clear, such as seasonal hurricanes or cyclones, or could be unanticipated such as power disruptions from a geomagnetic storm. Many hazards also occur irregularly such as a volcanic eruption or earthquake, which could have minor to major impacts. In working through a hazard identification, planners will want to gather information on the full range of hazards which should include:

- *Natural hazards* that occur because of nature's intersection with people and places. These might include floods, tornadoes, hurricanes, volcanic eruptions, earthquakes, and heavy rain or wind.
- *Climate change* certainly threatens businesses in locations from coastal zones experiencing a rise in sea-level to areas that experience higher temperatures resulting in crop damage, drought, or wildfires. Impacts to agriculture, tourism, and local businesses can be significant.
- *Hazardous materials* accidents can occur within a given industry, from a nearby industry, or from a transportation or site-specific spill. Even the BP oil spill in the Gulf of Mexico affected people along the coastline and inland when the seafood industry was compromised. Spills that happen on a major transportation artery can cause shelter-in-place orders or an expedited evacuation. A major accident, such as the Bhopal, India tragedy, can claim lives and affect hundreds of nearby businesses and their employees.
- *Transportation arteries and infrastructure* can be sources of disasters, for example, when trucking accidents lead to hazardous spills or an airplane crash causes damage, death, and disruption. A major bridge failure or train accident can derail employee travel patterns or the delivery of goods and services. If a public transportation system fails, which happened in the 2003 blackout in New York City, employees may have to stay at work or walk home. The cause was a problem hundreds of miles away in another state, where a problem with switching loads caused a failure with rippling effects. In 2018, a bridge failed in Genoa, Italy. Heavily traveled, the immediate impacts caused 43 deaths and subsequently disrupted traffic including business travel for some time.
- *Terrorism* can occur anywhere and anytime as can active attackers who invade a workplace. Assessing the history of terror attacks at similar industries could serve as a starting point to determine potential impacts and thus reveal possible strategies to protect intrusion or impacts. Terrorism should expand beyond the kind that people can directly inflict to include cyberterrorism, malware, and ransomware which have disrupted health care industries, election processes, and banking in recent years.
- *Workplace violence* can also occur anywhere. These can be motivated by any number of factors include potentially disgruntled employees or situations involving interpersonal disagreements or domestic violence. The U.S Bureau of Labor Statistics (2019) reports that in 2018 757 people died in the workplace because of intentional injuries by others.
- *Pandemics* have occurred throughout history, with the worst occurring in 1918 and 2020 worldwide. Yet, isolated outbreaks can also prove hazardous to businesses and employees. The 2003 SARS epidemic hit the Toronto, Canada area hard and served as a call to action. Nearly 800 died nationally, with 8000 cases spanning 26 countries. Though SARS has not

returned, extensive pandemic planning resulted which prompted early alerts for COVID-19. Nonetheless, COVID-19 resulted in hundreds of thousands of deaths worldwide.

- *Space weather* includes solar flares, geomagnetic storms, and related effects. Before thinking this seems far-fetched, consider that such weather can disrupt navigational systems, cellular service, and abilities to communicate. If your business relies on any of these devices directly or indirectly (such as through air travel for business purposes), it would be wise to follow space weather forecasts as well as those for severe storms.

Thus, a hazard identification begins with looking for potential risks that might interrupt or undermine the viability of a business. A wide scan should be undertaken, using credible sources to identify and outline the potential area hazards. Do not worry at this point about the odds that a specific hazard will happen. Rather, focus on the hazards that might happen. You will sort out the probabilities specific to your location later. Make a list using the template in Box 3.2 to aid your efforts.

Sources of hazard information

Finding information to inform your hazard identification is easy, in fact, you may uncover too much information. To make your search manageable and useful, focus on the range of hazards in your geographic region as well as those that could be encountered due to travel and use of the Internet. Businesses that involve trucking, for example, will need to identify a wide-ranging set of hazards their drivers may encounter from winter blizzards to springtime floods. A local childcare, though, may need to only focus on what will happen or influence their immediate area, such as a tornado or a seasonal respiratory illness. Databases exist on most hazards (again, see Box 3.1) so starting there might prove expedient and informative. Consulting with area emergency managers, meteorologists, weather services, and agencies tasked with homeland security may also prove useful.

Businesses may have people available to gather hazards information, such as an internal emergency manager or public safety officer. Smaller businesses may benefit the most from area emergency managers or might want to call on student interns or faculty in regional colleges and universities. Older members of the community may also be able to offer unique perspectives, such as their memories of the Blizzard of 1978 or the terror attacks of September 11, 2001. While details should always be confirmed, their experiences with an event may shed light into how events affected people and places, because everyone experiences it differently. Because disasters are not equal opportunity events and tend to affect lower-income workers and families the most, their perspectives should be consulted (Thomas et al., 2018). Lower income workers

BOX 3.2 Hazard identification template (with sample illustrations)

The purpose of this template is to inspire planners to collect information on area hazards. Planners may need to add or delete hazards as appropriate. Examples are provided to guide planners in the kinds of content they should fill in the open spaces.

Hazard	Previous history in area	Employee impacts in area	Business impacts	Probability of reoccurrence	Low, medium or high probability event?	Information source
Natural disasters						
Flood	1931 Flood through CBD					
Hurricane/cyclone						
Flood						
Terrorism, cyberattacks, and active attackers						
Terrorism	Sept 11, 2001	3000 deaths	High	Periodic, unknown	Medium probability, high impact	
Active attacker						
Malware attack						
Pandemic/viral						
COVID-19	2020	23,573 deaths 890,000 ill	Extremely high	Unknown, likely seasonal	High probability, high impact	www.who.int
Influenza	Annual	Varies	Varies	Seasonal	High probability, low to medium impact	
Zika	Annual	Varies	Varies	Seasonal	Low probability, medium to high impact	
Climate change						
Drought						
Coastal flooding						
Wildfire						
Transportation accidents						
Highway pile-up						
Infrastructure failures						
Bridge failure						
Power failure						
Space weather						
Geomagnetic storm						
Solar flares						
Hazardous materials accidents						
Offshore oil spill						
Internal spill/explosion						

may be among the first to lose their jobs, as witnessed in the pandemic of 2020. Similarly, historic patterns of discrimination and prejudice have marginalized many population groups, with disproportionate consequences. Thus, woman- and minority-owned businesses tend to bear harder impacts with higher failure rates (Dahlhamer & Tierney, 1998; Marshall, Niehm, Sydnor, & Schrank, 2015).

In the United States, hazard information for political jurisdictions is often easily accessible as FEMA requires state, tribal, territorial, and local governments to develop and maintain (renew every 5 years) hazard mitigation plans as a condition for certain types of federal funding. Also, the FEMA website contains historical information related to disaster declarations by political jurisdiction that can be helpful for understanding localized hazards. In addition, U.S. jurisdictions engage the Threat Hazard Identification Risk Assessment (THIRA) process (Box 3.3) as part of the Department of Homeland Security funding processes. Some jurisdictions make this information public, others shield the information from public release due to the sensitive nature of the contents. Finally, the Emergency Planning and Community Right-to-Know Act (EPCRA) requires the development of Local Emergency Planning Committees (LEPCs) that both develop a response plan and provide information about chemicals in the community to citizens. LEPCs can be helpful sources of information on hazardous materials in a community.

How to write up a hazard identification

How do you write up a hazard identification? Your data-gathering will need to be compiled into a usable reference format for your team to review and analyze. Consider Box 3.2 as an efficient overview and template for your analysis. Parts of the template include identifying areas hazards, the previous history of such

---

BOX 3.3 Department of Homeland Security Threat and Hazard Identification Risk Assessment Process

In the U.S., the Department of Homeland Security (2018) *Comprehensive Preparedness Guide (CPG) 201: Threat and Hazard Identification and Risk Assessment (THIRA) and Stakeholder Preparedness Review (SPR)* provides guidance for strategic planning levels. The CPG-201 process tasks planners with addressing five essential questions. These questions may be useful in surfacing hazards, capabilities, and gaps that a business should address.

1. What do we need to prepare for?
2. What level of capability do we need to be prepared?
3. What are our current capabilities?
4. What gaps exist between the capabilities we need and the capabilities we currently have?
5. How can we address our capability gaps? (DHS, 2018, p. 8).

These questions inform investment priorities for local, state, and federal grant funding targeted at sustaining or building community capability. The assessment also identifies capabilities that potentially degraded due to any number of factors, such as equipment that is damaged or expired.

events in your area, prior employee impacts, and the ways in which the hazard affected area businesses. While you could write this up yourself through an individual effort, it might make sense to rely on business colleagues to share the task. For example, your local chamber of commerce or its equivalent could convene a series of business continuity planning sessions to provide an overview of how to write plans. Many entries in your Box 3.2 template will be the same across the community – so why not share the information and reduce the workload? You will also want to include the source of your information in case questions arise and so that you can go back to that source when you annually update your plan. How do you know which hazards you should address first? Making that decision emerges through the next step, which is a risk assessment.

Risk assessment

Now that you have completed a hazard identification, your next step is to conduct a risk assessment. Risk assessments determine which hazards are the most likely to impact an area, with what level of impact, and how businesses would want to prioritize action accordingly. For example, many areas may be subject to hurricane or cyclone hazards, but some areas will bear a higher likelihood of landfall, storm surge, and wind damage. Such a seasonal risk can be anticipated, but not the exact timing, severity or specific locations affected. As another example, volcanos will impact a focused geographic area but have worldwide consequences such as when the 2017 Iceland (Eyjafjallajökull) volcano disrupted airline travel. This Icelandic eruption had significant economic impact by disrupting business operations depending on air transportation for employees and products. Volcanos also erupt at unspecified times, often with minimal advance warning like in 1991 with Pinatubo in the Philippines and 1989 Mt. Redoubt in Alaska (Schneider et al., 2019). The eruptions could be short-lived or extend for many years as lava flows intrude into business and residential areas. For example, in 2018 lava flows from Hawaii's Mt, Kīlauea volcano over a 4-month period forever altered the island resulting in the destruction of 700 homes and numerous roadways. Risk assessment tasks teams with identifying the most common and significant risks and an appropriate response.

Understanding probabilities and historic trends

Mathematicians, statisticians, and risk experts calculate the probabilities of an event occurring based on its prior incidence, in short, its historic prevalence and trends over time. However, the probability of something happening is just that – a chance that it could occur within a given time. Probabilities do not guarantee that an event will happen. You could go to a gambling casino and have a probability of winning something, but the odds that you will hit the big one are much trickier – it could happen, but it very well might not. The same is true with disaster events such as earthquakes or floods. An earthquake occurs because of a fault that lies within the earth. The probability that the fault will move varies from low to high depending on the type and location of the fault – and its history. People in California wait for the "Big One" to happen, because of the significance of the San Andreas fault (Jones, 2018). And indeed, The Big One could happen as seismologists warn us – just not when, with what kinds of impacts, or exactly where. People living further inland, in the midwestern U.S., tend to not worry too much about earthquakes. Yet, the New Madrid fault could move causing massive damage to businesses. An earthquake that occurred in the same area in 1811 moved the Mississippi River (Johnston & Schweig, 1996; Quarantelli, 1985). A similar event today would cause catastrophic damage to this important waterway not to mention riparian systems that include agricultural areas and flow alongside large cities like St. Louis and Cincinnati. Nonetheless, seismic building codes have been created in many higher risk areas (particularly in California) along with extensive planning, education, and training, sometimes at significant expense. Potentially affected businesses and agencies must determine what they can and will do, coupled with their best understanding of the risk and the resources available to act. Those decisions emerged out of hazards identification and risk assessment efforts.

Similarly, each year hurricane and cyclone researchers and experts offer forecasts of the coming "season." Coastal and island communities pay attention, because of prior and recent experience. But inland areas often fail to recognize risk associated with that forecast which can include extensive amounts of rain that beleaguer their communities. Flooding likewise follows probabilities, with hydrologists describing prior floods as 100-, 500-, or even 1000-year events. What do these numbers mean? A 100-year event means that, in any given year, the flood has a 1 in 100 chance of being matched or exceeded in any given year, a calculation created by hydrologists to suit government needs (Ceres, Forest, & Keller, 2017; Holmes & Dinicola, 2010; Mayo, 2019). However, just because that 100-year event happened 2 years ago does not mean a community gets the next 98 years free of risk. The estimation is a probability of when such an event could occur based on its annual exceedance probability which can be influenced by climate change, how an area grows, and the area's ability to address such risks (e.g., see Gallina et al., 2016; Pregnolato, Ford, Glenis, Wilkinson, & Dawson, 2017). A business may not see such a flood for another 98 years, but it *could* happen again next year. Yet, probabilities can be useful. For example, given that the New Madrid fault moves about every 200 years, the area may be overdue for such a shift (Tuttle et al., 2002).

So, focus on this: it might be that an area has not flooded since 1913 (a 1000-year event), but when it did the effects were catastrophic (Hinds, 2013; Williams, 2013). If your hazard identification uncovers similar information, take time to learn about ways in which the community may have mitigated those prior impacts. Does a floodwall or levee system now protect the downtown? How recently was that system challenged by mother nature? What happened as a result? What degree of confidence do your local government officials and floodplain experts now have in existing mitigative protections for the next event? Probabilities may make us think it will not happen again, but we need to be sure – and to be ready.

Thus, risk assessment can be undertaken using such advanced statistical models for some hazards, but a simpler approach can suffice for many businesses. To launch a simple risk assessment, consider each hazard and discuss whether it is a low or high probability event, then enter that into your table as shown in the template. How do you decide if it is low, medium, or high probability? By paying attention to frequency and historic impact. Seasonal monsoons in southern Asia occur with regularity and produce expected flooding at anticipated times – thus, a high probability event. Determining the frequency and history of prior impacts in the area(s) of business location would help to determine if those impacts have historically produced low, medium, or high consequences. While such an evaluation may be a judgment call, you will soon discern which hazards present the most frequent event and can focus business continuity planning on events most likely to happen and most likely to have an impact. Should you ignore low probability, high impact events? No, but all businesses will have to make a choice about what they can afford to plan for and the costs that such an effort will require.

Crafting scenarios

Now that you have completed a hazard identification and have a feel for the associated risks, you will also be able to identify impacts that might face your business. In writing a BCP, these hazards and risks can be used to generate scenarios that prompt planning team discussion, surface areas of concern, and generate creative problem-solving. Should you generate a scenario for every hazard? The answer within emergency management is generally "no" because many hazards create the same kinds of consequences: displacement, disruption, downtime, supply chain problems, telecommuting, alterations to shift work, communication issues, payroll – and more. Emergency managers generally use an "all-hazards" approach to write most plans so that they emerge with a plan that efficiently and sufficiently covers a range of possible events and outcomes.

For example, severe weather of all kinds may cause the same problems that occur in a terror attack: warning people, caring for the injured, accounting for employees at the site of an attack, and communicating about the key next steps everyone should take to respond and recover. Generally, then, crafting one plan to cover multiple scenarios will suffice for most businesses most of the time. Those in areas of specific risk – such as an area subject to earthquakes (see Brown et al., 2019) and related tsunamis that also serve as a host site for a nuclear plant – will require additional levels of planning. Care must be taken also for personnel who might be exposed to significant risks from an explosion to an attack or a tornado strike. With the 2020 pandemic, essential workers came to include not only health care and first responders but also postal workers, delivery people, grocery store workers, meat packing personnel, and custodians. Dozens of these essential workers died and, though pinning down the exact source of infection can be difficult, deaths could be linked workplaces for those in health care, policing, firefighting, and paramedics/EMTs. For an example of some scenarios and what to talk about, see Box 3.4. Clearly, human losses represent the worst possible outcome but they are not the only losses that businesses should consider, a topic this chapter turns to next.

BOX 3.4 Holding discussions around a scenario

Based on your hazard identification and risk assessment, draft a paragraph to spark discussion within your business. It might look something like these:

Your hazard identification and risk assessment has resulted in determining that an area dam represents a considerable risk in case of extended rainfall. Should the dam fail, your business and those you rely on will likely have 3–5 ft of water inside the business for up to a week.

Your business relies on Internet connectivity with data storage for critical records with confidential data. Employees notice a slowness in their connectivity followed by a complete loss of connection and lack of access to critical records. A ransomware message arrives soon after, demanding $30,000 to release the computers back to the company.

Historically, ice storms have led to power failures across the region and make travel nearly impossible. Your business or agency needs to continue to function as it provides critical services to the region. Given an anticipated power loss of up to 1 week accompanied by transportation challenges to clients, patients, or customers, the ice storm represents a potentially crippling event.

In your deliberations as a planning team, talk about the following:

- How will your business and employees pivot to adapt to a new normal – which might mean displacement or downtime that will affect productivity and revenue generation?
- Where else could your employees work toward the critical functions that your business addresses?
- Where is the backup site for people to work?
- Does everyone have what they need to do their core work?
- If you lose a portion of your workforce, what is the personnel backup plan?
- Where have you backed up key records and resources so that you can continue to operate?
- Who are the essential employees that need to work toward reopening the business and what are their tasks?
- What is an alternate procedure to get employees to their worksites or to the clients, patients, or customers they need to serve?

Loss estimation

Loss estimation should reveal the economic and human costs of a hazard by considering potential losses to the contents, the structure, the business uses and functions, and impacts to the human resources who fuel production and services (Brookshire et al., 1997). Each will add up to revenue and personnel losses, as well as the potential for losses related to legal concerns and civil liabilities.

Kinds of losses to estimate

In the U.S., the Federal Emergency Management Agency (FEMA) has produced some useful tools to conduct a loss estimation (see https://www.fema.gov/media-library-data/20130726-1521-20490-4917/howto2.pdf, Last Accessed April 29, 2020). In a straightforward way, FEMA walks planners through estimating direct impacts from structure, content, use and function losses to calculate a potential total loss. Totaling these losses can result in an eye-opening spreadsheet that calculates financial impacts and catches the attention of a CEO whose support is needed by planners. The kinds of losses start with inventories and estimations of replacement costs. Potential losses that should be considered include:

- *Structure loss estimation.* Most businesses operate out of set facilities which could range from a home office to a massive physical plant across multiple international locations. Structures owned by a business should be inventoried for the cost of replacement. The original value of a building should not be included, because replacement costs often exceed the original value of a structure. Discussion should then inspire reviewing insurance coverage in case of a structural loss.
- *Content loss estimation.* A second loss that should be inventoried includes the contents of a structure or the tools, resources, and assets needed to sustain the operation. That could be the fabric supplies in a quilting shop, the cattle, feed, and machinery in a dairy operation, the computers and related software needed in a tax accounting service, the assembly line and software systems need to make vehicles, or the products, carts, and computerized checkout counters in a grocery or retail setting. These assets could be lost in a worst-case scenario with a direct hit by a tornado or feel insufficient in a pandemic. Though conducting such an inventory could be daunting given the size of an enterprise, units within the business could amass the data using existing computerized inventories and databases. Do not forget the routine but needed items like desks, chairs, computers, printers, office supplies and the other business-specific contents needed like books and journals for teachers, vehicles and mowers for a lawn service, or protective equipment for police, fire, and health care workers.

Every asset matters and might be lost, damaged, or inaccessible or unavailable in a disaster.

- *Use and function loss estimation.* What is the business used for? What would happen if your business had to relocate? What if the business cannot perform business as usual? While that may be impossible to consider, the reality is that a business may need to adapt to a new normal with new ways to deliver goods and services. Or, the business may need to function in a new location with a different space, borrowed assets, and new personnel. A coffee shop might need to go mobile or a restaurant might want to establish a catering service while their space is being repaired or rebuilt or if normal operating hours become compromised. What would your business do if it faced a displacement? Even a home-based business may need to find a new place from which to operate. What are the costs of displacement? Consider these as a starting point to add up potential losses:
  - Moving expenses to get the business into a new location.
  - Storage costs for inventory and assets that will be used once the business re-opens.
  - Security for the damaged business area/plant.
  - Rental costs of a new space, vehicles, and related assets essential to operating.
  - Costs to turn on utilities and restore Internet capacity.
  - Purchases related to restarting or adapting to the new site or new normal.
- *Human resources loss estimation.* People represent critical assets for a business because their labor enables enterprises to succeed and recover. Employees may also be absent as they recover from injuries, pick up the debris in their own homes, or care for family members. They may be unable to commute due to transportation impacts or have evacuated to a significant distance. Formal medical leaves, caregiver leave, sick leave, and similar benefits may allow for their temporary absence – but the business will still have to cover those functions. Businesses that are "one-deep" in critical positions may face more difficult situations, but temporary workers can be hired. It can be hard to know where to start with such an estimation. Think about a pandemic scenario, where a percentage of your staff could be absent. What would you do if 10% were out sick? What if it was more? How would you continue operating even in their absence – what is the cost to your business to do so? A call to a temporary agency, costs for advertising for new employees, or an estimation of the hourly cost for a worker should be added in. The expenses of an employee who remains on the payroll should be added into the human impacts part of a loss estimation, as well as carrying the costs of benefits.

- *Revenue/financial loss estimation.* Businesses need to know the costs of operations versus the revenue that they generate. The difference results in profit or loss. Knowing what a business makes on a daily, weekly, monthly, or annual basis can assist in loss estimation around the financial impacts of a disaster – which can be significant. Related, a business will need to know what its existing financial resources will afford in terms of a safety net. Can the business operate without generating revenue? Most cannot – and would need to know how long their cash reserves and assets will cover known and anticipated expenses before additional measures need to be taken including closures and layoffs.
- *Legal concerns and civil liabilities.* Though most business continuity plans do not address legal matters, it is wise to do so. This is particularly relevant if a disaster happens on a business property and is related to that business, like a chemical explosion or hazardous materials spill. Injuries to employees, customers, clients, and the community can result in significant legal cases and costs even when cases are ultimately dismissed.
- *Insurance.* The costs of a deductible for insurance should be reviewed as it can be significant. Does the business have sufficient cash or assets that can be used toward a deductible? Insurance policies also can be radically different in cost based upon exclusions within the policy. It is important to understand triggers and limitations of coverage as part of cost recovery. For example, small businesses in Frederick County, MD found varying exclusions in policy that they were unaware of before considering impacts of potential civil disorder during the 2012 G-8 Summit. These exclusions included business interruption insurance exclusions for result of area closures by government actions and exceptions for events that resulted in the activation of the National Guard (Landahl, 2013). It would also be wise to bear in mind that, in a major disaster, policies could increase after the impact (increasing costs to the business) or could be cancelled completely.

Broader community impacts

Though many business impact analyses (discussed below) do not include consideration of broader community impacts, it is advisable to do so. Communities rely on their local businesses to provide products, goods, and services and to generate revenue impacts for other businesses like gas stations, dry cleaners, childcare facilities, health care settings, grocery stores and more. Revenue generation also includes taxes that help schools and government offices. Businesses often donate services, goods, or funds to local charities and fundraisers. People in a community also hold allegiance to preferred restaurants, bars, theaters, and stores. Families delight in holding special events at such places or look forward to concerts, plays, and dances. Schools and universities also serve as gathering sites for public lectures, performances, and athletic events that serve as a strong core within any community. Businesses often serve as social gathering places where the ritual of a birthday party or retirement matters, because of their social attachments to place (Hummon, 1990).

Businesses also exist because of their connection to the community and the value they generate. Any business loss will affect the community, from a neighbor who loses their home because they cannot pay the rent or mortgage to the corporation that always donated to the annual domestic violence shelter fundraiser. The loss of any business thus diminishes the broader community in some way. In many locations, those businesses serve as a means of survival. Communities today remain economically segregated in many places, with challenges that arise during a disaster. The existing food desert in an impoverished neighborhood becomes exacerbated when the lone grocery store fails. If public or private transportation sources fail, people without cars cannot travel to medical appointments or to work.

The community and their ties to businesses can also become a resource in enabling a business to survive. After the 1989 Loma Prieta earthquake, residents used their own hands to move a favorite bookstore to a new, temporary location several blocks away. In the 2020 pandemic, patrons supported side-lined restaurants and bars by buying gift cards from closed restaurants and using curbside, takeout, and delivery services. Those customer-based responses resulted in innovative business adaptations and jobs in a newly bustling home-based delivery service and efforts that produced masks, face shields, and other personal protective equipment. The broader community embraced such adaptations and stepped up to support businesses made fragile by the pandemic.

Business impact analysis

The prior work your team has undertaken in hazard identification, risk assessment, and loss estimation should help to surface potential impacts (Păunescu, Popescu, & Blid, 2018). To work through what is called a business impact analysis (BIA), this section identifies circumstances and concerns that a planning team should consider. To start, each business must attend to its financial bottom line. Thus, revenue verses expenses should serve as a first focus of attention. The loss estimation will help with this, with an eye to potential impacts such as:

- Sales losses and disruptions that could occur, when business close, or supply lines become disrupted. Customers could also lose confidence in a product and how it is prepared. Such problems arise not only during an infectious disease outbreak, but during more unusual events like the "mad cow disease" that devastated the British cattle industry or a widespread contamination (Pennings, Wansink, & Meulenberg, 2002).

- Customer and client losses can also occur, often through both direct and indirect impacts. Direct impacts, of course, mean that the business may not be able to open and function. But indirect impacts also occur. After the 2001 Nisqually earthquake near Seattle, Washington, the central business district suffered losses when customer traffic slowed (Chang & Falit-Baiamonte, 2002). Getting customers to return to stores, use services, and patronize businesses can take some additional effort. The business impacts can include loss of revenue, a diminished customer base, and an increased cost for marketing and advertising to lure people back.
- Contract losses might also ensue when a disaster happens, and those contracts could be significantly far away from the business that is impacted. An earthquake in Japan, for example, can disrupt production worldwide as directly impacted businesses have to cancel contracts for goods and services.
- Production disruption can also impact a business when an assembly line slows due to decreased demand. When people affected in an area lose their paychecks or savings, ripple effects occur to area businesses. Traditional customers may reduce, delay, or cancel their purchases, orders, or service needs. Or, a disaster in one part of a plant or facility can impede productivity in another part. Managing these kinds of production disruptions will be necessary to sustain business operations and stay afloat.
- Availability of workers at any point in time can also affect a business and negatively impact its bottom line. Employees can be directly affected if a disaster hits their homes or neighborhoods and they cannot work. Policies that disallow employees to enter an area may also disrupt business operations, such as when the U.S. closed its borders for COVID-19 which prevented agricultural workers from harvesting seasonal crops. Massive amounts of crops spoiled, affecting others downstream including farmers' markets, grocery stores, and delivery systems. Similarly, COVID-19 outbreaks at meat packing plants closed facilities temporarily while workers struggled with the devastating impacts of the virus from hospitalizations to deaths.

It is also true, though, that disasters present opportunities. For the COVID-19 outbreak, the U.S. used the Defense Production Act to compel the auto industry to make ventilators. At the height of the outbreak in the U.S., 30 million people became unemployed when stay-at-home orders went into effect and only essential services remained open. Health care industries, though, as well as congregate care settings, grocery stores, and delivery services, managed to remain operational, to add new employees, and to adapt. After September 11, businesses moved into providing homeland security services in multiple nations, created new ways to repel attacks, and offered consulting services to reduce potential impacts.

Finally, the timing of a disaster can create a significant impact. Businesses often move through cycles when they are busier, such as accounting firms during tax season or hospitals during influenza season. The potential impacts could be significant if disruptions, downtime, and displacement occur during that time of high worker productivity – including mission fulfilment as well as revenue generation for the firm. For example, agricultural producers face seasonal threats annually, as they monitor weather for extended rainy periods during planting season or the impacts of extreme temperatures that threaten agricultural producers with late freezes, wet planting seasons, or droughts. Schools and universities also have high-impact times when they onboard new students, which can be disrupted by power outages, cyberattacks, or a pandemic. Tourist areas enjoy higher incomes during vacation season, where disasters can have more devastating impacts on bottom lines. Clearly, life cycle impacts can influence revenue for the year or over time, such as the debilitating consequences of extreme drought that lasts for years. Knowing whether and when to adapt matters, and can save the business, alter its direction, or cause its demise. Business impact analysis looks at what a business can tolerate until revenue and resources become so depleted that it reaches a point where it must adapt or perish. Each business will have to make its own decision about that point, which is determined by individual tolerance to risk and the willingness of the business and its employees to make changes, to withstand impacts, and to realize a need to move on.

Determining acceptable losses

The goal of a loss estimation is to amass the valued inventory and assets of a given enterprise. The cost to replace those losses, from minor impact to complete shutdown and displacement, will garner the attention of an executive no matter how far away their office is from the emergency planner and their team. Ultimately, any business will have to determine how much loss is acceptable before they have to make adaptations that will range from minor to major before they have to alter their services, reorganize their systems, change their workforce, shutter their doors, or face the challenges and responsibilities of reopening.

Now that you have completed your hazard identification, risk assessment, and loss estimation, your planning team will need to decide what level of risk your business is willing to take. Look at Box 3.1 again, where you focused your hazard identification and risk analysis. Most businesses will probably choose to plan for a mid-range level of disasters that occur regularly with some significant impacts. Why not plan for all eventualities? With terrorism as an example, it is easy to understand that a potential attack concerns all of us, though some areas appear to be more likely targets such as malls, heavily trafficked tourist areas, or symbolic sites. Most businesses will not face terrorism directly, so they might rank it as a low probability but high consequence event. Return on investment in terrorism protection activities is not easily measured, for example large investments in visible security measures at the Disney Springs shopping and entertainment venue area in Orlando, FL deterred an attacker who did surveillance at the location, later choosing the Pulse nightclub as the target of the attack (Hesterman, 2020). Worldwide, flooding is the most common hazard that occurs and can range from minor to catastrophic. Based on the geographic regions for your business, which hazards should you focus on? Most businesses will have to focus on the ones that occur most frequently and with moderate impacts.

Why plan for these kinds of events – the ones in the mid-range? Why not plan for the catastrophic events? Because most businesses – indeed most governments – lack the resources to do so. While it would be wonderful if a levee system could protect a city of 500,000 with the highest level of protection, its local government probably cannot afford such a means of protection. Hurricane Katrina pushed a category 5 storm surge into the City of New Orleans in 2005. Once rebuilt, the levee system returned to its pre-storm category 3 level of protection, albeit having fixed several pre-Katrina levee weaknesses. Why? Because building to a level 5 was not affordable. Businesses will have to make similar judgment calls on what they can and cannot feasibly undertake. We all must make decisions on the risk level we can tolerate.

Such discussions about the limits to risk reduction can be difficult, even wrenching. Countries worldwide struggled with risk reduction decisions as the COVID-19 pandemic evolved – to stay closed, adapt, or reopen. Leaving work or stay at home situations could potentially further imperil worker and community health, inundate health care facilities, and cause an additional wave of illness and death. Job losses had particularly affected lower income workers, who also faced worry and fear as facilities re-opened with the virus still prevalent. Businesses had to face the limits of their rainy-day funds and decide between protecting the public health and staying afloat. Deciding what to do, how to do it, and how to pay for it comes down to choices about publicly shared values and personal commitment to the public good while facing the loss of one's future. It is not an easy situation to be in, and the conundrum is not unique to pandemics as such discussions take place around wildfires, flood risks, and cyber/terror attacks. What is essential in such discussions is straightforward acceptance that risk will occur, and that each business has to make a decision that works for them within the limits they will face. Making those decisions can become clearer after moving through a next essential action that examines potential losses.

Mitigating losses

Strategies for businesses to sort through once determining loss estimation begin with the idea of "mitigation." This word means that businesses will take actions to reduce their potential losses (Schwab, Sandler, & Brower, 2016). While a full scope of mitigative actions for all businesses lies outside the scope of this volume, a planning team can start by looking at what might be lost and designing strategies to reduce such impacts (see Box 3.4). Costs for each can vary from inexpensive to very costly. Each business will have to decide what they can afford vis-à-vis the risks they perceive to exist around them. Two main kinds of mitigation strategies usually go into place: structural and non-structural mitigations.

Structural mitigations refer to the built environment, such as blast-resistant windows in an area subject to projectile damage from high winds, tornadoes, hurricanes, or cyclones. Structural mitigations can be located within a business, like a safe room, or can come from a shared risk reduction process across the entire community, like a levee system along a river. A workplace with concerns for active attackers or terrorist threats can install curbside bollards that deter intrusion from a vehicle carrying explosives. Areas subject to flood risks can join with city or community efforts to strengthen levee systems and stockpile sandbags, develop relationships with disaster recovery services that remove water, and build new sites outside a floodplain or elevate a business to withstand intrusion. Working with the broader community to put structural mitigation measures into place makes for excellent use of business owners' time and taxes, because an effort put into the common good provides protection for the business, those that rely on the business, and the broader community. Sharing the risk through such joint efforts and funding also reduces the financial impact on any single business. Businesses can and should become champions for local mitigation projects for just this reason.

Non-structural mitigation efforts refer generally to those that are not "built" like insurance, planning, education, and policies. Business continuity plans serve as a perfect example of non-structural mitigation because of the effort that goes in to thinking about how a business will need to recover from a disaster. Workplaces that already have telecommuting policies in place will be more ready to move their human resources and needed computing assets to a new site – and to get back up and running remotely. Educational campaigns help people understand potential hazards and their risks as well as how the business will protect employees and operations, respond in a crisis, and recover from an impact. Businesses will also have to make informed choices about the financial resources they will want to put into place before a disaster. Insurance serves as one means to reduce the financial impacts of a disaster, though care must be taken to cover as many losses as possible. The business will have to decide how much coverage they want to secure, for what items, and at what level of cost. Insurance policies may need to be written carefully to cover specific losses unique to the business, which can prove costly. Consideration must also be given to what policies will not cover, including specific kinds of hazards like flooding, high wind, terrorism, and pandemics. Though business interruption insurance can be purchased, the cost may not be feasible vis-à-vis the risk the hazard presents. Businesses should also anticipate that even insurance companies can fail financially or may cancel their policies for specific risks. An earthquake swarm, for example (e.g., such as what was caused by fracking in Oklahoma) meant that home-owners and businesses could not buy earthquake insurance to mitigate their risks. Policy and deductible costs also increased.

Businesses should also work at conserving cash in a rainy-day fund. In higher education, universities try to have 90 days of cash on hand to cover specific expenses, in part to earn their Moody's ranking as an indicator of financial stability. The 2020 pandemic pressed in hard on universities worldwide, and Moody's moved higher education from the category of "stable" to "negative." Some colleges furloughed or laid off employees when their cash in hand diminished and enrollment dropped. Financial advisors often recommend that home-owners have 3–6 months of cash or liquidable assets to cover expenses. Those 3–6 months serve as a cushion during which a business can regroup and begin recovery. To calculate that amount, determine monthly expenses then multiply by the number of months. For example, monthly expenses that total $10,000 should result in a target of $30,000 (minimum) to at least $60,000 set aside in a rainy-day fund to withstand a disaster.

Essential actions

In this chapter, planners and their teams learned about essential actions that need to be taken to be ready to write a business continuity plan. Those essential actions include:

- Conducting a hazard identification to surface area threats.
- Assessing those risks to determine their frequency and history and to prioritize the ones most likely to have at least a moderate impact.
- Conducting a loss estimation to see what the business can withstand and what could be lost.
- Thinking through a business impact analysis based on the first three essential actions.
- Deciding what losses will be acceptable and identifying a means to reduce those losses.

References

Brookshire, D. S., Chang, S. E., Cochrane, H., Olson, R. A., Rose, A., & Steenson, J. (1997). Direct and indirect economic losses from earthquake damage. *Earthquake Spectra, 13*(4), 683–701.

Brown, C., McDonald, G., Uma, S. R., Smith, N., Sadashiva, V., Buxton, R., … Daly, M. (2019). From physical disruption to community impact: Modelling a Wellington Fault earthquake. *Australasian Journal of Disaster and Trauma Studies, 23*(2), 65–75.

Ceres, R. L., Jr., Forest, C. E., & Keller, K. (2017). The 100-year flood seems to be changing. Can we really tell? In: *AGUFM, 2017, NH34B-03.*

Chang, S. E., & Falit-Baiamonte, A. (2002). Disaster vulnerability of businesses in the 2001 Nisqually earthquake. *Global Environmental Change, Part B: Environmental Hazards, 4*(2), 59–71.

Chartres, N., Bero, L. A., & Norris, S. L. (2019). A review of methods used for hazard identification and risk assessment of environmental hazards. *Environment International, 123,* 231–239.

Dahlhamer, J. M., & Tierney, K. J. (1998). Rebounding from disruptive events: Business recovery following the Northridge earthquake. *Sociological Spectrum, 18*(2), 121–141.

Gallina, V., Torresan, S., Critto, A., Sperotto, A., Glade, T., & Marcomini, A. (2016). A review of multi-risk methodologies for natural hazards: Consequences and challenges for a climate change impact assessment. *Journal of Environmental Management, 168,* 123–132.

Hesterman, J. (2020). Terrorism: What protection officers need to know. In S. J. Davies, & L. J. Fennelly (Eds.), *The professional protection officer: Practical security strategies and emerging trends.* Cambridge, MA: Butterworth-Heinemann.

Hinds, C. C. (2013). *Columbus and the great flood of 1913: The disaster that reshaped the Ohio Valley.* History Press.

Holmes, R. R., Jr., & Dinicola, K. (2010). *100-Year flood—It's all about chance.* U.S. Geological Survey General Information Product 106.. 1 p. Available at (2010). *https://pubs.usgs.gov/gip/106/.* (Last Accessed May 4, 2020).

Hummon, D. M. (1990). *Commonplaces: Community ideology and identity in American culture.* Suny Press.

Johnston, A. C., & Schweig, E. S. (1996). The enigma of the New Madrid earthquakes of 1811–1812. *Annual Review of Earth and Planetary Sciences, 24*(1), 339–384.

Jones, L. (2018). *The Big Ones: How natural disasters have shaped us (and what we can do about them).* Doubleday.

Josephson, A., Schrank, H., & Marshall, M. (2017). Assessing preparedness of small businesses for hurricane disasters: Analysis of pre-disaster owner, business and location characteristics. *International Journal of Disaster Risk Reduction, 23,* 25–35.

Kato, M., & Charoenrat, T. (2018). Business continuity management of small and medium sized enterprises: Evidence from Thailand. *International Journal of Disaster Risk Reduction, 27,* 577–587.

Landahl, M. R. (2013). Businesses and international security events: Case study of the 2012 G8 summit in Frederick County, Maryland. *Journal of Homeland Security and Emergency Management, 10,* 1–21.

Marshall, M. I., Niehm, L. S., Sydnor, S. B., & Schrank, H. L. (2015). Predicting small business demise after a natural disaster: An analysis of pre-existing conditions. *Natural Hazards, 79*(1), 331–354.

Mayo, T. L. (2019). Predicting the 100-year flood to improve hurricane storm surge resilience. *Notices of the American Mathematical Society, 66*(2).

Păunescu, C., Popescu, M. C., & Blid, L. (2018). Business impact analysis for business continuity: Evidence from Romanian enterprises on critical functions. *Management & Marketing Challenges for the Knowledge Society, 13*(3), 1035–1050.

Pennings, J. M., Wansink, B., & Meulenberg, M. T. (2002). A note on modeling consumer reactions to a crisis: The case of the mad cow disease. *International Journal of Research in Marketing, 19*(1), 91–100.

Pregnolato, M., Ford, A., Glenis, V., Wilkinson, S., & Dawson, R. (2017). Impact of climate change on disruption to urban transport networks from pluvial flooding. *Journal of Infrastructure Systems, 23*(4)04017015.

Quarantelli, E. L. (1985). What is disaster? The need for clarification in definition and conceptualization in research. *Disasters and Mental Health: Selected, 10*, 41–73.

Schneider, D. J., Guffanti, M., Lisk, I., Mastin, L. G., Pavolonis, M., Sr., Tupper, A. C., … Rennie, G. (2019). Thirty years of explosive volcanism and aviation: Redoubt, Pinatubo, Eyjafjallajökull and beyond. In: *AGUFM, 2019, V33A-07.*

Schwab, A. K., Sandler, D., & Brower, D. J. (2016). *Hazard mitigation and preparedness: An introductory text for emergency management and planning professionals.* CRC Press.

Thomas, K. L., Wilson, T., Crowley, K., Hughes, M., Davies, T., Jack, H., … Leonard, G. (2018). *An example of how community participation can be successfully incorporated into the disaster risk assessment process, Aotearoa-New Zealand. .*

Tuttle, M. P., Schweig, E. S., Sims, J. D., Lafferty, R. H., Wolf, L. W., & Haynes, M. L. (2002). The earthquake potential of the New Madrid seismic zone. *Bulletin of the Seismological Society of America, 92*(6), 2080–2089.

U.S. Bureau of Labor Statistics (2019). *Fatal occupational injuries for selected events or exposures. Retrieved from (2019). https://www.bls.gov/news.release/cfoi.t02.htm. (Last Accessed June 15, 2020).*

U.S. Department of Homeland Security (2018). *Comprehensive Preparedness Guide (CPG) 201: Threat and Hazard Identification and Risk Assessment (THIRA) and Stakeholder Preparedness Review (SPR).* Washington, DC: Government Printing Office.

Williams, G. (2013). *Washed away: How the Great Flood of 1913, America's most widespread natural disaster, terrorized a nation and changed it forever.* NY: Pegasus Books.

CHAPTER 4

Parts of a business continuity plan

Scenarios and decision-making

Business continuity planners must now take their hazard identification and risk assessment into consideration, which may involve addressing a range of potential disasters (see Chapter 3). However, planning for everything will result in a complicated set of plans that may prove overwhelming to complete. In most emergency management agencies, those involved create plans with the understanding that many functions remain the same despite the type of disaster. We call this the all hazards approach. For example, cyberattacks, natural disasters, and hazardous materials accidents all generate communications challenges, so designing a robust means to maintain communications internally and externally spans many types of disasters. Similarly, business continuity planners will have to deal with potential supply chain disruptions, human resource impacts, displacements, and disruptions to their operations and more – whether facing a hurricane or a pandemic. Thus, it makes sense to plan generally rather than specifically, especially for those just getting started with BCP. In this section, we will look at what various kinds of disasters can generate and the disruptions they cause. But keep an eye on the similarities across the events, which should drive subsequent planning efforts and eventually work their way into your planning software (see Box 4.1).

Pandemic

In late 2019, a new or "novel" illness arising out a "coronavirus" family (eventually designated as COVID-19) appeared in the People's Republic of China. The World Health Organization (WHO) first reported learning of the illness on 31 December 2019. Declaring a "global health emergency" by January 30, 2020, WHO implored all nations worldwide to help stem the rapid spread of the illness which had affected over a dozen countries. One day later, the U.S. declared a public health emergency.

Business Continuity Planning. https://doi.org/10.1016/B978-0-12-813844-1.00005-1

BOX 4.1 Resources for business continuity planning

The Federal Emergency Management Agency in the U.S. offers a free, downloadable software suite that provides training tools, forms, videos, handouts, brochures, and a planning software program at https://www.ready.gov/business-continuity-planning-suite (Last Accessed 2/4/2020, original software loaded in 2016). Called the Business Continuity Planning Suite, the materials include a software generator that creates a usable plan with your input. Materials walk users from understanding the definition of business continuity planning to testing the plan in a six-step process. Smaller businesses and agencies can also download shorter templates that walk planners through the basics sufficient to cover their concerns.

Additional software plans can be purchased from an array of vendors. What should you look for in a vendor software?

- *Ease of use*. Given that most businesses lack expertise in business continuity planning or disasters that might befall them, any software should be user-friendly, and menu driven. Categories should make clear sense and should enable planners to work steadily through the various elements to generate a usable plan. Aspects of the plan should also include guidance on the various terms and sections so that planners can make sense of them.
- *Essential elements*. Business continuity planning relies on several common elements. Software should address, at a minimum:
  - *Critical functions.*
  - *Key people and teams and their roles.*
  - *Backup materials needed to continue operations and how they may be accessed.*
  - *A communication protocol and process.*
  - *Consideration of relevant scenarios.*
  - *The role of key resources such as technology.*
  - *An inventory of essential items needed.*
  - *List of resources, vendors where essential items can be secured.*
  - *Emergency contact information for teammates.*
- *Final product*. While electronic access to the plan is certainly essential, so is a hard copy or pdf of the plan so that it can be shared among a wide array of users and also stored in multiple locations – yes, back up your business continuity plan.

WHO and affected nations moved quickly to stem the epidemic. China shut down most travel into and out of Hubei Province and built temporary hospitals for thousands of patients, followed by required quarantines and arrests for noncompliance. By the end of January, countries like the U.S. had urged international travelers to avoid flying to China. Planes evacuated Americans who went through a quarantine procedure after landing to ensure the virus had not been transmitted. By the end of February, 2718 confirmed deaths had occurred in China with another 44 outside of the affected nation (World Health Organization, 2020). The World Health Organization ranked nations from 1 to 3 with various precautions. China and South Korea received a level 3 indicating that people should avoid nonessential travel to these nations. Japan was designated a level 2 requiring enhanced precautions and closed its schools for the month of March. Italy followed soon after, with U.S. schools closest to outbreak areas also closing. Universities recalled students studying abroad and cancelled travel to specific high-risk areas. The pandemic spread further and faster than the public and many workplaces anticipated. The virus overwhelmed hospitals and health care providers desperately tried to save lives in nation after nation, with insufficient resources. Affected nations closed their borders and mega-cities issued stay-at-home orders that lasted from weeks to months. Workplaces worldwide closed and economies suffered as millions of people moved on to the unemployment rolls. It felt overwhelming, everywhere.

Pandemics represent one of the top concerns of business continuity planners (Droogers et al., 2016). In the case of the novel coronavirus, stock exchanges reacted negatively, beginning in Asia. Companies and educational institutions stopped travel for business and research ventures. Tourism, also a significant industry, dropped with heavy impacts to airlines, hotels, and conferences. Several cruise ships fell into quarantine status, with thousands of passengers and employees contracting the virus. Collateral damage ensued, which disrupted business transactions and effects to world trade. Stock markets reacted, with the U.S. Stock Exchange dropping over 12% in 1 week at the end of February, the worst decline since the recession of 2008. Early, estimated global impacts in stocks reached three trillion USD by the end of February, prompting G7 countries to act (CBS, 2020). Airline industries suffered with flights into and out of "Level Three" countries (with the highest number of cases and deaths) discontinued followed by rapid and sustained declines in passenger numbers. The United Nations directed funds into highly vulnerable and lesser developed nations and global banking systems prepared to do the same (U.S. Department of the Treasury, 2020). In early March, the World Health Organization again urged all governments to take immediate action to contain the worldwide public health emergency.

Schools, colleges, and universities reviewed their pandemic plans. Efforts began to focus increasingly on higher risk locations. In the state of Washington in the U.S., multiple nursing home residents remained under quarantine, an early indicator of what would happen nationally with possibly 30,000 or more nursing home residents perishing from the virus. Nineteen firefighters who responded to the nursing home fell under subsequent quarantine, reducing available emergency personnel to care for those affected in a pattern that soon spread across the U.S. (Gunderson, 2020). Health care workers also faced significant exposure as did police officers and paramedics, with hundreds losing their lives.

Business continuity planners often rely on scenarios, not unlike the coronavirus pandemic, to engage planning teams in making critical decisions. In the case of a pandemic – which many businesses had failed to prepare for – what might a business have to consider? Think through these questions with a pandemic scenario but remember that these questions could also pertain to a natural disaster, terror attack, or hazardous materials accident:

- If officials asked you to close your workplace, how would you continue to operate?

- Can you close or is your workplace essential to the public good such as a police or fire station?
- Is it possible to operate your business, or portions of your business, through an online capacity? Have you prepared for such a scenario with access to critical records and resources through a cloud technology? The 2020 pandemic resulted in millions of people trying to work from home, pushing Internet capacities everywhere and revealing weaknesses with infrastructure needed for businesses to survive. Those who could not telecommute or did not have such an option often lost their jobs with millions applying for unemployment assistance and economic business loans and grants. A second wave of economic impacts occurred again as reopening efforts had to slow when the virus surged again, followed by furloughs and layoffs at closed businesses and then at universities and colleges.
- How many people can work from home to sustain business operations? In 2020, workplaces scrambled to find sufficient resources to support home-based personnel and IT offices became strained with trying to provide technology, update software, teach people to connect remotely, troubleshoot from a distance, and create new online websites and platforms to adapt their business. Other settings that rely on technology, such as vehicle assembly lines, stopped production, and laid off workers.

- How would your business handle employees out sick or, in the case of a natural disaster, out with an injury or needing to clean up from the flood or tornado? Some industries, such as a hospital or nursing home, must remain operational. Health care workers, emergency responders, and emergency managers may have to address significant issues such as:
  - Can you release or relocate patients to another facility? In 2020, nursing homes and congregate care sites became high risk locations and established early quarantines to save lives.
  - Is quarantine mandated? What are the impacts? With quarantine or stay at home orders, travel became restricted and commerce slowed. Worksites deemed "essential" by their governments had to provide security passes to enable travel, with stay at home the best strategy to slow the pandemic. Employees included in quarantine settings became anxious and overworked. Some committed suicide.
  - Can you cancel elective surgeries or similar procedures? In 2020, health care facilities cancelled or delayed such elective procedures until the pandemic had passed. The economic hit of this action became significant in marginalized areas, particularly rural hospitals already struggling to survive.
  - How many personnel are required to keep essential operations functioning? What happens when they become exposed to the virus and need to be quarantined? What kinds of personal protective equipment

(PPE) must be offered to reduce their exposure and keep the business functional? How does a business provide a safe working environment when reopening becomes possible?

  o Do you have a mutual aid agreement with another, similar business, or a contract with an area temporary employment agency in case personnel become compromised? As the 2020 pandemic developed, calls came for retired military and health care workers to return to work. Given their potential high-risk categories by age, they often supported non-pandemic cases including through newly evolved telehealth options. The mutual aid agreement worked, and health care adapted – two important strategies to keep in mind for your own planning even if you are not in a health care setting.

• What if your business remained open, but a key partner in another location closed? What would be the impacts on your productivity, operations, inventory, and workflow?

  o For a health care setting, for example, how would your business secure products if those items are shipped in or purchased from the affected area? Governments had to provide resources by opening stockpiles (if they had them) and using military units to deliver desperately needed resources, set up field hospitals, or establish virus testing operations and food delivery. Some businesses, like grocery stores, limited what a customer could buy which impacted highly vulnerable populations particularly hard due to their social isolation. Other agencies stepped up and adapted by focusing on socially isolated and high-risk populations and providing useful services.

In short, what do you have to do to stay in business? Such activities organize around what emergency managers call critical functions, which serve as the core activity in planning for business continuity. In recovering from or adapting to a disaster, emergency, or pandemic what really must be done?

Technology disruptions

Cyberattacks and technology disruptions have increased in recent years, with disruptions significant enough to halt city operations, stop air travel, and interrupt surgical procedures. In 2018, the city of Atlanta endured a lengthy cyberattack. The event began with computer outages, most likely caused by a ransomware attack where cyber criminals demand payment (usually in bitcoins) in exchange for releasing computer access (Shamma, 2018). The situation was serious, with the large city's data held hostage. Business continuity efforts, to address critical functions, began to unfold. Police officers wrote out reports and tickets by hand, instead of using their vehicles' on board computerized systems. Other affected departments, representing nearly half of the city's services, included corrections and court systems, water, human resources, parks, and planning (Deere, 2018a). Operations had to revert to backup procedures from the Thursday when the attack began until the following Tuesday, though effects continued for several weeks. While the large, international Atlanta airport remained unaffected, they did shut down the airport's wireless connection which caused delays with business communications.

Immediate costs reached an estimated $2.6 million, for digital forensics, crisis communications assistance, incident response consulting, and hiring additional staff (Newman, 2018). Some estimates placed the cost of the total event at $17 million USD (Deere, 2018b). Though the city did not pay the ransom, they did push financial resources into upgraded security and purchased over $1 million in new hardware more resistant to cyberattacks. A few months after the attack, authorities charged two Iranians with disrupting computers tied to city operations. Ransomware attacks also target small businesses, shutting down their computers and access to online orders, databases, and software needed to operate. Being ready if such an attack occurs, and preferably putting measures in place to prevent or divert such an attack, can help a business to survive. What would you do if you lost computers, Internet connectivity, and the ability to use phones?

Natural disasters

Earthquakes occur worldwide and with catastrophic results at times. The 2004 Indian Ocean earthquake generated a tsunami that killed approximately 300,000 people and devastated independent fishing operators, sizable merchant sectors, and tourist economies in 13 nations. Estimated economic costs reached billions of USD between 1995 and 2015 with 25% of those losses caused by floods (UNISDR, 2015). Far more common than a tsunami, floods remain the most common disaster worldwide, representing nearly half of all disasters and with economic costs reaching staggering totals (UNISDR, 2015). It does not take a hurricane or cyclone to cause such damage, either. In 2019, tropical storm Imelda dropped so much rainfall in Texas that economic costs surpassed $5 billion USD (NOAA, 2019). The costliest tornado occurred in Joplin, Missouri (U.S.) and exceeded $2.8 billion in damage to a local hospital, multiple schools, hundreds of homes, and dozens of businesses (NOAA, 2015).

What must you do when a disaster strikes, whether it is a pandemic, a cyberattack, or a natural disaster? The answer lies in what a business identifies as its critical functions, which reflect the core operations of a business and enable work teams to focus on what must be done, when, and by whom after something terrible happens.

Critical functions

Writing a set of critical functions defines the first step in crafting a business continuity plan. Critical functions reveal what the business must do to keep working, creating, supplying, or supporting its customers, clients, patients, students, and partners (see Box 4.2). A bakery, for example, will need to maintain its supply chain to produce wedding cakes, cater events, and sell daily baked goods to the public. They will also need to keep sufficient staff on hand to continue production, market their items, take orders, and deliver products. In an emergency, trying to do everything may be very difficult – but focusing on critical functions can help the business to survive.

Some planners make the mistake of listing job responsibilities or functions which can bog down a planning effort. In identifying critical functions, planners should avoid generating a massive list of all the company's functions to focus on the core, essential duties. In a way, writing critical functions is like what happens when medical professionals do triage – what must be done first, second, and third to save a life – in this case, to save a business? A hospital needs to keep the power on to save and care for patients. A university needs to keep teaching to support its students. An accounting firm needs to continue audits, completing taxes, and serving clients. At the heart of a critical function lies the mission of the business, agency, or organization. Thus, a good starting place would be to look at the company mission – what must be done to support that mission?

BOX 4.2 Critical functions

Consider critical functions for these business types, which might embrace the following:

- Restaurant
  - Reduce food spoilage by maintaining refrigeration.
  - Restore utilities and/or use generators to continue functioning.
  - Purchase essential supplies for food preparation and serving.
  - Communicate with staff and public about open hours.
  - Prepare, serve, and/or deliver food safely.
- Higher education
  - Continue delivering critical courses essential for students to make degree progress.
  - Communicate with faculty, staff, and students about situation, safety, and restoration of operations.
  - Protect and preserve faculty research, from data storage to plant and animal species, biological specimens and organisms, and perishable cultures.
  - Insure that students who live on campus have adequate shelter, food, and sanitation.

- Safeguard critical records and engage in alternative ways to manage data depending on the event (e.g., IT failure).
- Move courses online or on to campus as needed.
- Assisted living center.
  - Maintain safety and health of residents.
  - Communicate with staff, residents, and family.
  - Support facility with adequate heat, cooling, water, and sanitation.
  - Transport residents as needed to alternate sites for safety and health care.
  - Establish self-isolation or quarantine as needed and set up means for residents to communicated with loved ones.
  - Insure proper levels of staffing to support residents including staff transportation.
  - Coordinate with appropriate agencies and partners depending on event (e.g., evacuation, pandemic/quarantine, power outage).

Another way to surface critical functions is think through what a company would need to do in one of the scenarios that started this chapter. For example, what are the critical functions for a bakery to remain operational if a quarantine is imposed? A critical function might focus on "continue baking" (see Box 4.3). To maintain the critical function of producing goods, the bakery might lay out how to adapt for takeout, curbside pickup, or home delivery. By identifying critical functions, a company decides in advance what they intend to do and works through direct and indirect impacts, downtime, and displacement decisions. For example, an earthquake might undermine the utilities needed by the bakery, thus prompting a potential relocation for temporary operations. Or, they might decide to narrow their production to specific items that could generate income. After hurricane Katrina, for example, some New Orleans area bakeries focused on re-opening for the critical Mardi Gras season and the production of traditional king cakes. The decision generated a nice economic boost after 6 months of closure, which eager customers embraced as the heart of their culture.

Writing critical functions should involve appropriate personnel and units to yield the best insights and ideas. Let us continue with the example of critical functions in the smaller bakery. The critical function of "continue baking" would benefit from the eyes and minds of multiple workers who fill the variety of roles needed in bakery production. Their insights will be crucial because they will see the production process from various viewpoints. Sometimes, planners make the mistake of trying to do the work of writing the entire plan or hiring consultants to do so. Writing critical functions thus depends on empowering employees to speak out, so that people with key knowledge will inform the plan. Everyone should have a voice, from the custodian who must clean a

BOX 4.3 Examples of critical functions

Accordingly, the bakery's critical functions might look like this, despite the type of disaster:

- Keep baking:
  - Identify essential workers/team to keep on production line.
  - Identify backup and temporary workers.
  - Identify alternative vendors for key ingredients and resources (e.g., packaging containers).
  - Identify alternative location if main production line is affected and establish agreement for use.
- Market and sell products to maintain revenue:
  - Include traditional and expand on social media.
  - Pivot to curbside, takeout, and delivery options.

- Maintain operational capability and building use:
  - Keep utilities on including generators as necessary through pre-identified maintenance team.
  - Maintain vehicle fleet including backup source for delivery, procuring ingredients, and employee transportation.
  - Maintain supply chain with alternate vendors for key ingredients.
  - Continue with cleaning protocol per local health department requirements.
  - Identify backup workers and supervisors.

production line to the bakers who know machinery operations to management responsible for insuring that a product leaves the line and gets to the store shelves.

It is worthwhile to mention planning scale here. Small bakers, for example, might be able to craft a BCP that is relatively focused. But a large bakery factory that produces goods for mass distribution might need a far wider set of plans. Each of the above critical functions, then, might evolve a more detailed plan with additional critical functions. Let us take the example of a university. While the university should have a plan for what it will do to complete its instructional mission, individual units might need separate plans as well. Each of those plans would need to spell out critical functions from units that cover human resources, facilities, tutoring services, admissions, advising, housing, dining, student financial aid, the health and wellness center, and more. Thus, a decision will need to be made about how many plans a business needs based on the size of the enterprise. Generally, the larger the enterprise, the more plans will be needed. Those plans will need to then be collated so that each unit knows what is happening above, below, and within the organizational layers of the business. Management will ultimately have responsibility to know what is happening within and below their units so that surprises do not occur in an emergency. It will not work to assume that another unit will procure the needed resources for your unit – you must be sure. Management needs to be sure they agree with any anticipated expenditures, such as what would need to be procured to pivot from restaurant service to takeout or from in person instruction to online delivery. Both management and workers need to agree on what comes first as well.

Prioritizing critical functions

Additional steps will need to be taken once the team has generated the key critical functions. To start, which functions are the most important? For example, if a tornado has affected your business (even indirectly), marketing may need to be delayed while the business redirects available employees (whose homes were not affected) to set up generators, deliver existing orders, and secure supplies to maintain production. A second step after generating a list of potential critical functions, then, is to establish their priority to the company – so that leaders, managers, and employees will know what will be tended to first. Top critical functions should be sorted for life safety (e.g., the critical function of communicating a tornado warning) followed by mission critical functions specific to the various units within the business (e.g., continue teaching classes, produce food, deliver payroll). Thus, writing critical functions spells out what needs to be done, the order in which the business wants it to be done (which may need to flex in an actual crisis), who will make sure it gets done, and how they will do that essential work. Critical functions represent the core agency activities that must continue for the business to remain viable (Whitworth, 2006).

Who bears responsibility for those critical functions? To whom is something like communication or marketing assigned? What are the first steps they should take? This is where the diversity of the planning team matters because their eyes and minds will help to inform essential steps and develop work teams (discussed below). For a restaurant, it may be essential to have someone start a generator immediately after a tornado or power outage to prevent food spoilage. By designating a "restore power" critical function and assigning to a work team, businesses can ensure that highly prioritized functions like turning on the generators will be undertaken immediately. Refrigerators and freezers often contain food worth thousands of dollars that a business owner invested in – and needs to keep in business.

Thus, to ensure that the critical function is launched per the planned protocol, and with flexibility built in for unexpected twists, turns, and new events, teams should be developed and trained to respond. Those teams should be prepared to not only implement the plan but to be flexible, because disasters challenge even the best plans and the most experienced professionals. In the 2020 pandemic, universities shifted online which required teams with distance education expertise to guide more traditional, face-to-face teaching faculty to deliver their courses. Dog trainers moved to virtual sessions. Psychologists and physicians delivered services via telehealth platforms. Orchestra members livestreamed across continents to create classical and jazz pieces. Grocery stores and pharmacies offered special hours for high risk populations to decrease their potential exposure. Convention centers transformed into makeshift hospitals and tents went up in public parks to accommodate the ill, supported by humanitarian groups and the military. Tens of thousands of retired health care providers volunteered. Virtual meeting platforms emerged as a key tool to facilitate meetings of emergency teams in workplaces everywhere. Television and news programs produced broadcasts from people's basements. Even retail stores pivoted to curbside or delivery services and began to promote "Work at Home" clothing lines.

Finally, identifying critical functions suggests an immediacy and a need to act. When a team establishes the critical functions that enable them to meet the mission and intent of a business, they will also have to create a time frame in which those functions should be restored. In a major disaster, you cannot do everything so what comes first? Re-establishing IT so that employers and employees can communicate? Enacting policy changes to support employees or how things are handled? Each critical function should carry with it a preferred time frame (hours, days, weeks) for re-establishment so that teams can prioritize which function they need to take on first. In the pandemic of 2020, IT teams emerged as a critical function. Within IT, they had to manage continued support of and maintenance on existing systems while also transitioning employees to work from home. Even tech support changed, and eventually had to move to virtual operations to protect their personnel and maintain their critical functions. What made that work so well, in so many cases, was the work team that had been created to handle the critical function of IT support. In short, how much downtime can your business tolerate in its respective critical functions before assigned work teams make that function operational again?

Work teams for critical functions

Each critical function should have designated personnel and/or teams who can step in and lead efforts to restore the critical function (see Box 4.4). They should train and practice before the event and attempt to anticipate all eventualities (see Chapter 7 for ideas on how to do this). Teams should also identify backups or successors because the anticipated leadership or personnel might not be available. For COVID-19, health care workers fell ill from caring for their patients. Backups came from across affected nations and sometimes internationally, as when Chinese physicians went to Italian hospitals to help. Identifying backups may mean the difference between a business surviving or failing and whether it can support its functions as critical as patient care.

Planners should go beyond just identifying a backup and consider what that backup needs to know and do. For example, has sufficient cross-training

---

BOX 4.4 Critical functions and work teams

Critical function: Converting courses for online delivery

- Workshop team to organize and deliver webinars training faculty and staff to use online platforms and best practices for online courses.
- Mentor team to work one-on-one with those new to online delivery.
- Identify and share best practices for unique teaching including dance, chemistry, statistics labs.

Critical function: Continue to produce and sell menu items at our restaurant (while keeping existing personnel employed albeit in new roles)

- Menu Selection Team, which downsizes menu using available items for adaptation to takeout, curbside pickup, or delivery.

- Procurement Team, which assesses the availability of ingredients and sources them for use by the Cooking Team including personal protective equipment.
- Marketing Team, which creates advertising at reasonable cost to alert customers to new modes of delivery for the restaurant and the safe ways in which food is produced and made available.
- Online Team, which takes orders via phone, messaging, online system and relays information to the cooking team and coordinates with delivery services.
- Cooking Team, which produces the food via incoming orders and prepares them for packing suitable for takeout, curbside pickup, or delivery.

occurred so that the communication liaison backup knows how to talk and what to say to the public, or to families, or to students? Or, in a school or university setting, who will step into a classroom to carry on the critical function of instruction? Do they have the expertise they need as well as access to course materials? By ensuring that qualified and prepared backup people and processes are in place, an enterprise increases its chances of success. Thus, businesses should follow the strategy of "next person up" as The Ohio State University football team did in their 2014 season – when they lost top two quarterbacks. The third string quarterback was selected Most Valuable Player when the buckeyes won the National Championship, in only the second game he ever started.

The culture of a workplace, region, or nation should also be considered when designing work teams (Warrick, 2017). Cultures reflect values that influence behavior, like the buckeye "brotherhood" that led an unknown quarterback to not let down his team. Workplace culture influences how leaders make decisions, how teams come together, and how employees behave. The military, for example, relies on a hierarchically driven command system which enables them to efficiently train, deploy, and use people and assets in high stress environments. Other workplaces follow a different set of norms. Silicon Valley, for example, includes numerous businesses that rely on flexible, less hierarchical cultures to encourage innovation and speed (Steiber, 2018). It might not be unusual to see an employee skateboarding through the workplace to encourage self-expression. Cultural expectations thus set out how work teams will be organized, interact, and perform. Planners should take some time to talk about their workplace culture and how it may influence critical function work teams. Will they be loyal? Innovative? Will they be collaborative, or will they require a chain of command? The workplace culture will set the tone for how critical functions play out post-disaster.

Critical function work teams can be organized in several ways. In a disaster, a hierarchical system may not be as effective as hoped, because disasters always present new challenges that require flexibility, creativity, and improvisation (Kendra & Wachtendorf, 2003; Neal & Phillips, 1995; Seifert, 2003; Zimmerman, 2003). Many businesses already expect that critical thinking, problem-solving, and collaboration is essential in personnel (Lieberman & Parker, 2018). Indeed, businesses rely on such personnel, like a restaurant manager who serves a wide range of palates and nutritional situations or an accountant who sorts through widely varying circumstances during tax season.

To form work teams for critical functions, planners must:

- Think about the organization's culture and structure. How does the business expect people to behave? Those behavioral expectations will influence how the work team will operate in a crisis.

- Find the best people to shoulder a task and lead a work team through a potentially stressful situation with a heavy set of responsibilities. Finding the best people may mean looking carefully for people who can work through challenges with innovation and creativity.
- Form work teams that will interact well together as they learn and then launch the critical function actions.
- Identify backups and successors for each team member in the event they are unable to support the critical function. Not everyone is in town or at work when the disaster happens, as they can be on business travel, vacation, or out sick. Workers also bear responsibilities for family members as caregivers and parents and may have their own homes impacted (see Chapter 6).
- Create and test a means to contact people in an emergency and to enable them to reach the area for their critical function via technology (work remotely), transportation (be on site), or both. Call down lists serve this purpose as do email and text groups.
- Train and exercise the team specific to their critical function (see Chapter 7).
- Determine how the team will be scaled up or down and at which decision points that will occur. People will need rest when working in an emergency. A catastrophic event will require people beyond the initial team envisioned in a plan. Teams need to be on shifts that include breaks for psychological respite and physical recovery from related exhaustion.
- Sustain the team. Besides rest, people will need protective equipment, resources, food, hydration, showers, clean clothes, time with their families, and both physical and mental health breaks including days off and entire vacations. When planning the team, incorporate each of these elements into how to sustain the team over days, weeks, or even months. People may want to work 24/7 but such a strategy will not succeed, and the team will break down.
- Cross train the team to do each other's critical function tasks. What happens if a key person is unavailable – who will do their work? Someone will have to step in and by cross-training in advance, time and lives will be saved.
- Make sure that your work teams will have access to the resources that they need to restore and support their critical function.

Assets and workarounds

To be sure your teams will have what they need, several kinds of asset inventories must take place while writing a business continuity plan. As addressed in

Chapter 3, one type of inventory totals up key assets of value to the business, thus you can look back to the loss estimation to start this activity. Do you have what you need to sustain the business, specifically for each critical function? Is it an assembly line with computerized sections and machinery or a theater space with costumes and props? The assembly line may require an IT work team that has backup parts and programs to restore operations after a power loss or cyber-attack. An acting troupe may need mobile wardrobe and make up kits so they can produce Shakespeare from the park instead of the theater. Whatever makes your business move along and provide goods, services, and resources to customers, clients, and audiences should be inventoried so that you can identify, pre-disaster, where you need to have resources not only for daily use but for an emergency.

What do you use in the business daily from the early arriving janitor to the late-night analyst? Here is where the diversity of the planning team matters once again. Even a short inventory will reveal the company's assets and enable thinking through what should be kept on hand or in an offsite storage area to reduce the risk of total loss. If a tornado strikes a primary business, will offsite storage contain enough materials to become operational again? Or does the business rely on vendors to provide key resources? Who is the backup if the vendor cannot supply what is needed? In the 2020 pandemic, when hand sanitizer ran out globally, the alcohol beverage industry adapted to provide this product to entire cities and health care units. Where will you secure what your work teams will need and what is the backup plan?

What does the workforce need to keep operating? Businesses rely on general software as common as word document or presentation software or as specialized as software or platforms used by a dentist to care for patients, by retail stores to sell and manage deliveries, or by publishing companies to create and edit newspapers and books. Being able to use and restore those software packages requires knowing how to do so, preferably with an IT team, but potentially as the sole owner of a small business. Having a software list with instructions on how to reload, restore, or reboot reminds business owners to keep it backed up and ready, and with vendors in mind to provide support. Having backups of key documents, preferably in a secure offsite location, will also be crucial. Does everyone backup their work on a regular basis?

Next, what does the workforce need to know how to do? Can everyone remote in from home? Do they all have the computers and connections to make that happen? An asset inventory thus also includes buildings, furniture, special equipment, hardware, software, office supplies, assembly line materials, spaces, and persons needed to meet its mission both onsite and potentially offsite due to temporary or permanent displacement. Such an asset inventory can be as simple as listing the number, types, and locations of computers or as complex as a system-wide compilation of current inventory kept up in realtime. If people must work remotely, you will want to know what they will need and where your assets have gone. For those essential workers who remain, how will they be resupplied as needed? How will shifts be covered and with what priorities for work? Who will be their backups? Critical function planning answers these questions.

Communication tools should also be inventoried. Though cell phones have become far more robust, they have failed in certain disasters and will certainly fail given extended power outages. Therefore, an asset inventory will include identifying an alternate means to communicate. After the 1992 hurricane Andrew knocked out power in southern Florida for nearly a month (and batteries failed in radios), local governments lofted color-coded balloons to indicate shelter, food, and first aid stations. It may have been old school, but it worked. Hopefully, you will have other alternatives, but you never know so it is best to plan for multiple techniques including those that do not rely on power. Likewise, units that rely on power will need alternatives and workarounds. Could you imagine a significant power outage or loss of Internet while trying to register thousands of students at fall orientation? Most students are only there for the day of orientation, so you might miss the chance to electronically register them for classes. What are your workarounds, from how you prefer to communicate to what you may have to do in this or a similar situation?

Asset identification should also consider transportation and special equipment. If the 3D printer is a key resource, include it in an inventory with a plan to protect it and back it up. If your workforce lies in an area subject to heavy snowfall, you may have (or need) a specialized transportation system. For example, the state of Minnesota relies on volunteers with Hummers to transport health care workers to hospitals during severe winter weather. A business vehicle fleet may include buses, rental cars, delivery trucks, front loaders and skids, or snowmobiles, boats, and planes. Fleet inventory will need to consider supportive resources and equipment as well – gas and oil, certainly, but also mechanics, specialized sites to manage the fleet, and a work team dedicated to keeping vehicles moving. What kinds of backups could be used, such as an understanding with a similar business, local school, or public transportation system to use their vehicles?

Finally, assets include alternate locations where a business or service can relocate even if temporarily. What critical functions can be shifted to alternate locations? If a space is lost, like a dance studio that floods, where can that business re-open? You may need a formal or informal agreement to use an alternate space. Communities may also work collaboratively to set up temporary space to keep businesses afloat and tax revenue coming in. For the Loma Prieta earthquake of 1989, Santa Cruz downtown businesses relocated into a Tent Pavilion.

Though downsized in space, retailers could remain open and earn a living. After the tsunami in 2004, Oxfam funded the restoration of an entire commercial sector in Vailankanni, India. Banks offered micro-loans to smaller, often woman-owned businesses. The pandemic of 2020 required that additional buildings and hospital tents be erected as well as the use of military hospital ships. The 30-plus years between Loma Prieta and COVID-19 suggest that displacement will always need to be considered. Thus, planners working through critical functions should identify alternate spaces suitable for a new location, establish formal MOUs, and update them regularly. As one example, the American Red Cross creates formal MOUs with local sites to serve as temporary disaster shelters. That agreement lays out various responsibilities from who to contact to how sheltering activities are paid for and who cleans up the site afterwards.

One last lesson comes from the pandemic of 2020. Within weeks of the pandemic entering the United States, large industries shut down including car manufacturers. The Defense Production Act was used to require some manufacturers to support health care needs. Other industries stepped up independently. Battelle, located in Ohio, leveraged its systems to decontaminate N95 masks, which have a high level of protection against viruses, for re-use. Given critical shortages of the masks, the technology used proved invaluable to the health care industry when the technology cleaned up to 80,000 masks daily. Each mask could be sanitized up to 20 times, extending use through a life-saving transformation of the industry's potential (Hale, 2020). Battelle proved to be a most useful vendor and asset to the health care sector, an example of an essential partnership that saved lives and met a critical functional need. Being able to pivot and adapt – matters at both the business and humanitarian levels.

Essential partnerships (vendors, contractors, consultants, officials)

The next step is to consider your critical functions and who you rely on externally to keep your business operations going. Depending on one's business, a wide array of external partners may become essential to economic survival. Planners should consider integrating their most likely partners, including business partners and vendors, utility companies, internet and phone service providers, elected officials, and emergency managers, into their BCP. Such interorganizational collaboration and coordination represent crucial acts to take in advance of a disaster (Comfort & Kapucu, 2006).

On September 11, 2001, many businesses suffered heavy damage levels both inside and outside of the World Trade Centers in New York City and in the

Pentagon. Of crucial importance, the city's emergency operating center (EOC) had been in the World Trade Center. Business partners became crucial to their survival. Emergency managers, who evacuated from the building, regrouped and rebuilt the EOC in a new location within days. Vendors proved particularly important, supplying cell phones and cellular towers, power and cables, chairs and tables, computers, printers, food, and more (Kendra & Wachtendorf, 2003). The EOC teams were able to function again because the businesses and communities around them stepped up and provided critical supplies essential to this massive emergency.

Considering the resilience demonstrated in New York, emergency managers and public safety personnel should not be overlooked as potential partners. Ideally, you and your planning team have already met your local emergency managers during your hazard identification (see Chapter 3). If not, pick up the phone and call them now. They will be well-versed in what will happen during an emergency and how important matters may play out. If you lose your utilities, emergency managers will know how utilities will prioritize area restoration. Areas that include hospitals, for example, usually have a higher priority. If hospital care is your business, knowing the priority may be helpful to determine when or if you need to get additional fuel for generators or evacuate patients to another facility. By making a connection with area emergency managers, planners can learn such crucial information in advance of a disaster – because those same emergency managers will be too busy during an event.

Key business partners should also be consulted prior to a disaster happening, as they will prove useful to keeping your business – and theirs – going. If you are a grocery store owner, relying on key vendors and local producers will prove critical to not only you, but your broader community. The asset inventory that you just completed should reveal potential concerns in your supply chain or should prompt conversations with those vendors about what they will do in an emergency. If you know their plans to remain operational, you can improve your plans and anticipate disruptions – and how to address them. Business partners might include contractors and consultants, from those who provide specialized services to those who have been identified for outsourcing. Their availability may vary after a disaster, so once again having a conversation with them about their own emergency plans and how they plan to handle continued support should take place during this part of the business continuity planning process. You will want to keep those key ingredients coming to sustain bakery operations or continue personal protective equipment (PPE) for a health care setting, nursing home, or school in a pandemic.

Elected officials may also prove useful in an emergency. Typically, elected officials and government employees address mass emergencies through enacting or changing policies, funding new initiatives, serving as advocates, and negotiating resources for a stricken area. They also provide leadership, stay in touch with their constituents, offer information and guidance, and convene groups to work on the disaster. In a large enough disaster, they may organize collective efforts to restart the economy by alerting businesses to resources, opportunities, and workshops that get them moving again. By participating in civic efforts prior to a disaster, a business will get to know the government officials and employees and know who to turn to in a disaster. Government officials will also get to know the businesses in their region better and who they can call on to help the community and/or to help directly with aid and programs.

A typical strategy used to maintain effective partnerships, other than contractual obligations, is the development of a Memorandum of Understanding/ Agreement or a Mutual Aid Agreement. Common elements of these written and signed agreements include (for a sample, see https://emilms.fema.gov/ IS706/assets/WyomingTemplate.pdf, Last Accessed March 31, 2020):

- The parties and their capabilities.
- An agreement on what kind of aid will be rendered.
- Steps on how to request aid.
- Communication protocols.
- Personnel assignment and supervision.
- Equipment, use, and repairs/support.
- Covering costs and reimbursement procedures.
- Insurance, indemnification, liability, relevant laws, and related matters.
- Restrictions under certain conditions, such as a global pandemic and conditions under which an MOU cannot be expected to be fulfilled.

Essential actions

- Identify 4–6 critical functions for your business continuity plan if you are a small business. If you are a larger business, create planning groups that link to critical areas and have them craft their own plan with multiple, appropriate critical functions. Then, link those plans together with an overarching framework to ensure that everyone is on the same page.
- Develop work teams and their backups that will take on each critical function. Begin to think about training them, which will be more fully developed in Chapter 7.
- Inventory key assets that will be needed by the work teams in addressing their critical functions. Which internal employees or teams and/or vendors and their backups, link to and/or support those key assets?
- Lay out the key partnerships that you need to ensure that each critical function can be undertaken successfully: vendors, emergency managers,

other agencies and businesses, government officials, and even the broader community. Host conversations to understand each other's availability and business continuity plan so that you know who you can rely on.

- Identify alternate sites where your critical functions can set up in an extended displacement. Develop formal MOUs with those locations now.

References

CBS News (2020). *G7 finance chiefs vow to "use all appropriate policy tools" to avert virus "risks". Available at (2020).* https://www.cbsnews.com/live-updates/coronavirus-outbreak-death-toll-us-infections-latest-news-updates-2020-03-03/. *(Last Accessed March 3, 2020).*

Comfort, L., & Kapucu, N. (2006). Inter-organizational coordination in extreme events: The World Trade Center attacks, September 11, 2001. *Natural Hazards, 39,* 309–327.

Deere, S. (2018a). Atlanta city employees turn on computers for first time since hack. *Atlanta Journal Constitution,.* March 27, 2018.

Deere, S. (2018b). Confidential report. *Atlanta Journal Constitution,.* August 1, 2018.

Droogers, M., et al. (2016). European pandemic influenza preparedness planning: A review of national plans, July 2016. *Disaster Medicine and Public Health Preparedness, 13*(3), 582–592.

Gunderson, L. (2020). *19 Washington firefighters quarantined after coronavirus exposure at Kirkland nursing home. Available at(2020).* https://www.oregonlive.com/coronavirus/2020/03/19-washington-firefighters-quarantined-after-coronavirus-exposure-at-kirkland-nursing-home.html. *(Last Accessed March 3, 2020).*

Hale, C. (2020). *Battelle deploys decontamination system for reusing N95 masks. Available at(2020).* https://www.battelle.org/newsroom/news-details/battelle-deploys-decontamination-system-for-reusing-n95-masks. *(Last Accessed March 31, 2020).*

Kendra, J. M., & Wachtendorf, T. (2003). Elements of resilience after the World Trade Center disaster: Reconstituting New York City's Emergency Operations Centre. *Disasters, 27*(1), 37–53.

Lieberman, I., & Parker, M. (2018). The value of the arts within a liberal arts education: Skills for the workplace and the world. In *Exploring, experiencing, and envisioning integration in US arts education* (pp. 125–141). Cham: Palgrave Macmillan.

Neal, D. M., & Phillips, B. D. (1995). Effective emergency management: Reconsidering the bureaucratic approach. *Disasters, 19*(4), 327–337.

Newman, L. (2018). Atlanta spent $2.6M to recover from a $52,000 ransomware scare. *Wired,.* April 23, 2018.

NOAA (2015). *The 10 costliest U.S. tornadoes since 1950...in event-year and 2015 dollars. Available at (2015).* https://www.spc.noaa.gov/faq/tornado/damage$.htm. *(Last Accessed February 12, 2020).*

NOAA (2019). *U.S. billion dollar weather and climate disasters:* (pp. 1980–2019). *. Available at(2019).* https://www.ncdc.noaa.gov/billions/. (Last Accessed February 12, 2020).

Seifert, J. (2003). The effects of September 11, 2001, terrorist attacks on public and private information infrastructures: A preliminary assessment of lessons learned. *Government Information Quarterly, 19,* 225–242.

Shamma, T. (2018). Atlanta paralyzed for more than a week by cyber attack. *NPO Newscast (Ailsa Chang, host),.* March 30, 2014.

Steiber, A. (2018). Management characteristics of top innovators in silicon valley. In *Management in the digital age* (pp. 23–43). Cham: Springer.

U.S. Department of the Treasury (2020). *Statement of G7 finance ministers and central bank governors. Available at(2020).* https://home.treasury.gov/news/press-releases/sm927. *(Last Accessed March 3, 2020).*

UNISDR (2015). The human cost of weather related disasters *1995-2015.* United Nations International Strategy for Disaster Reduction: Centre for Research on the Epidemiology of Disasters.

Warrick, D. D. (2017). What leaders need to know about organizational culture. *Business Horizons, 60*(3), 395–404.

Whitworth, P. M. (2006). Continuity of operations plans: Maintaining essential agency functions when disaster strikes. *Journal of Park and Recreation Administration, 24,* 4.

World Health Organization (2020). *Coronavirus disease (COVID-19) outbreak. Available at(2020).* https://www.who.int/emergencies/diseases/novel-coronavirus-2019. *(Last Accessed February 27, 2020).*

Zimmerman, R. (2003). Public infrastructure service flexibility for response and recovery in the September 11th, 2001 attacks at the World Trade Center. In J. Monday (Ed.), *Beyond September 11th: An account of post-disaster research* (pp. 241–268). Boulder, CO: Natural Hazards Research and Applications Information Center.

CHAPTER 5

Planning for disruptions

Introduction

Earlier chapters in this text examined the process of identifying and analyzing potential hazards to facilities, personnel, and processes. This chapter focuses on various disruptions caused by failures within critical infrastructures that support businesses. The following sections of this chapter will examine issues related to disruptions across several infrastructure components including utilities (electric power, water, and natural gas), transportation systems, supply chains, communications, and information technology that have been concern to businesses for some time (Nigg, 1995; Tierney, 1997). Because critical infrastructures have become highly interdependent, outages in one system often affect other systems such as when electrical power outages disrupt computer use, cell phone service, and water pumping stations (without backups). Once the chapter introduces readers to possible disruptions, ways to manage related downtime and displacement are introduced followed by the ways in which the critical function component of a BCP connects to infrastructure failures and outages.

Infrastructure disruptions

Infrastructure disruptions represent potentially costly impacts for businesses, and related restoration of lifeline services influence and may determine the quality of business recovery (Chang, 2016). The U.S. Department of Homeland Security under *Presidential Policy Directive 21 (PPD-21): Critical Infrastructure Security and Resilience* designated 16 sectors of critical infrastructure (Obama, 2013). Under PPD-21, each sector is assigned a federal Sector-Specific Agency (SSA) that serves to coordinate information sharing, protective guidance, and response actions across the sector (see Box 5.1). In short, critical infrastructure agencies shore up businesses through offering widespread support and protection to reduce business disruption. Such critical infrastructure represents lifelines for businesses of all kinds that rely on utilities, Internet, and

Business Continuity Planning. https://doi.org/10.1016/B978-0-12-813844-1.00003-8

BOX 5.1 U.S. critical infrastructure sectors and agencies.

In the U.S., critical infrastructures link specific agencies to support a widespread set of sectors. To illustrate, the U.S. Department of Homeland Security supports sectors that include chemicals, commercial facilities, communications, critical manufacturing, dams, emergency services, government facilities, information technology, nuclear reactors and transportation systems with support from additional agencies as warranted. As examples, the Department of Defense provides sector support for the Defense Industrial Base, The Department of Energy for the energy sector, and both Health and Human and Services (HHS) and the Department of Agriculture for food-related sectors. HHS also supports the health care sector while the Environmental Protection Agency supports water and wastewater.

The purpose of tasking these agencies with support is because the named sectors represent crucial assets to the American people and the economy. Within the U.S., private sector provides the bulk of the nation's energy infrastructure including power, oil, and gas essential to businesses. Similarly, the communications sector enables businesses to connect using planetary, satellite, and wireless connectivity. Commercial facilities include eight subsectors: entertainment and media, gaming, lodging, outdoor events, public assembly areas like convention centers and zoos, real estate, retail, and sports leagues. Resources in support of security and related infrastructures for the commercial sector can be found in the links below.

Resources, last accessed July 27, 2020: *U.S. Department of Homeland Security (DHS), https://www.cisa.gov/critical-infrastructure-sectors; Commercial Sector resources, https://www.cisa.gov/cisa/commercial-facilities-resources.*

transportation systems to generate revenue, pay employees, and stimulate the economy. Many countries have undertaken similar lifeline vulnerability assessment to address potential risks. The New Zealand Lifelines Council, for example, has examined points of concern in the nation's supply chain. Failures could prove catastrophic not only with loss of service but impacts on business viability. For New Zealand, those include a refinery responsible for about 70% of the nation's fuel, an oil depot and pipeline, a gas field that supplies the bulk of the two islands' natural gas, a telecommunications exchange, and various ports including the main gateway airport in Auckland where 75% of international visitors arrive (New Zealand Lifelines Council, 2020). In short, national investments in critical infrastructure represent a crucial source of support for businesses and the nation's economy.

In the U.S., each sector and some sub-sectors maintain Information Sharing and Analysis Centers (ISACs). These centers facilitate information sharing among infrastructure owners and operators to assist in the protection of facilities, personnel, and systems from physical threats, cyber events, and other hazards. The ISACs function to collect, analyze, and distribute information on threats and hazards to sector partners. The products provide members with information to mitigate risks and enhance resilience of facilities and systems. This kind of coordination represents a crucial way to support the interdependent nature of critical infrastructure elements. In New Zealand, for example, electricity supports the functional capacity of other lifelines. Any failure means that other systems become more important, such as fuel delivery for generators, that must be transmitted via transportation systems (New Zealand Lifelines Council, 2020).

To understand the interconnectivity, consider how a derecho damaged parts of the U.S. in 2012, causing cascading effects within the energy sector that spread outward to other systems. A derecho is a widespread and persistent straight-line windstorm associated with fast moving showers or thunderstorms including dangerous bow echoes, squall lines, and straight-line winds (U.S. Department of Commerce, 2018). The 2012 derecho traveled east-southeast starting in the U.S Midwest, affecting Indiana in the early afternoon and proceeding through the Mid-Atlantic region around midnight. The storms impacted 10 U.S. states (Indiana, Ohio, Kentucky, Pennsylvania, West Virginia, Maryland, Virginia, Delaware, New Jersey, and North Carolina) and the District of Columbia. Winds pushed forward at a speed of 60 mph in most locations, with some winds above 80 mph and a few pockets above 100 mph (hurricanes start at 75 mph, see U.S. Department of Commerce, 2013). The derecho resulted in widespread damage and power outages lasting up to 7 days. Power failures caused cascading effects across other critical infrastructure including transportation, communications, public health, oil and natural gas, water, emergency services, and government facilities. Effects included transmission line damage, cell tower and switching equipment failures, problems with water pumping stations resulting in boil water orders, loss of services at banks, gas station pumps, and traffic signals (among many), and heat related deaths. The event also compromised abilities to use emergency services including 9-1-1 call management (National Infrastructure Advisory Council, 2013) (Box 5.2). Power loss represents one of the most impactful critical infrastructure disruptions that can be experienced by businesses (Graveline & Gremont, 2017). Widespread infrastructure issues from the derecho event, Superstorm Sandy, and multiple catastrophic storms in 2017 led FEMA to protect critical infrastructure as "the most fundamental services in the community that, when stabilized, enable all other aspects of society to function" (FEMA, 2019, p. 9).

Critical infrastructure has become so important that it now lies embedded deeply in the United States' *National Response Framework* (DHS, 2019b). The lifelines concept has motivated "infrastructure owners and operators, and other partners to analyze the root cause of an incident impact and then prioritize and deploy resources to effectively stabilize the lifeline" (DHS, 2019b, p. 8). In short, nations and critical infrastructure companies have moved to support economic and residential sectors through anticipating and preparing for disruptions that will harm businesses. The good news is that critical infrastructure specialists work hard, often out of view, to ensure that businesses will continue to operate in a crisis. While challenges continue to appear, a responsive network of agencies across many nations tries to protect those who invest their resources in business enterprises.

BOX 5.2 The National Infrastructure Advisory Council (NIAC) Resilience Framework.

The National Infrastructure Advisory Council (NIAC) Resilience Framework consists of three key components. The first is the pre-event robustness of infrastructure systems with ability to absorb impacts from incident. The second is the resourcefulness during the event to manage incident impacts as they develop. Third, is the rapid recovery of systems after an event. The goal is to ensure that a system becomes sufficiently capable that it rebounds when an impact occurs. The U.S. power grid, normally highly reliable, has introduced redundancy into its systems to increase robustness. That is a good thing, because American reliance on electricity has increased 5 times what it was just 50 years ago (NAIC, 2010). While normally robust, a blackout occurred in 2003 caused by human and computer errors. The event affected 50 million people in 8 states and 2 Canadian provinces. Outages lasted from a few hours to several days with some rotating blackouts occurring for up to 2 weeks (NAIC, 2010). Commuters in New York City walked miles from work to their homes, while businesses shut down. One study found that mortality and respiratory hospitalizations increased two- to eightfold over other similar hot days (Lin, Fletcher, Luo, Chinery, & Hwang, 2011) with significant economic losses and repair costs (Goodrich, 2005). When hurricane Maria damaged Puerto Rico, military units provided fuel to support hospital generators (see Fig. 5.1).

The Department of Homeland Security (2019c, p. 11) defines resilience as the "ability to prepare for and adapt to changing conditions. This means being able to withstand and recover rapidly from disruptions, deliberate attacks, accidents, or naturally occurring threats or incidents. Resilient infrastructure must also be robust, agile, and adaptable." Supporting investments to make utilities more resilient is a sound idea, as one study found that if a terror attack compromised electrical power in Los Angeles, the business interruption costs would total $20.8 billion USD without prior interventions. With resilience adjustments, the costs are estimated to drop to $2.8 though the study did not estimate other utilities or deaths from power losses (Rose, Oladosu, & Liao, 2007).

Sources: *Goodrich, J. N. (2005). The big American blackout of 2003: A record of the events and impacts on USA travel and tourism.* Journal of Travel & Tourism Marketing, 18(2), 31–37; Lin, S., Fletcher, B. A., Luo, M., Chinery, R., & Hwang, S. A. (2011). Health impact in New York City during the Northeastern blackout of 2003. Public Health Reports, 126(3), 384–393; National Infrastructure Advisory Council (2010). A framework for establishing critical infrastructure resilience goals. *Available at https://www.cisa.gov/sites/default/files/publications/niac-framework-establishing-resilience-goals-final-report-10-19-10-508. pdf (Last Accessed July 26, 2020); Rose, A., Oladosu, G., & Liao, S. Y. (2007). Business interruption impacts of a terrorist attack on the electric power system of Los Angeles: Customer resilience to a total blackout.* Risk Analysis: An International Journal, 27(3), 513–531.

Anticipating disruptions

While broadscale disruption to infrastructures as daily occurrences remain rare in many nations, businesses must plan for them due to the damaging consequences that may result. Disruptions could arise as part of an approaching hurricane (when you have some time to prepare) or in a rapid onset event that typically lacks a warning, such as a technological failure, hazardous materials accident, or earthquake. Events with warning periods should allow time for adaptive response actions (Faupel, Kelley, & Petee, 1992). In this case, preparatory actions will have been useful. Critical function teams will have time to ready the business for any anticipated impact hours, days, or even weeks ahead of an anticipated event. Events without warnings represent more significant challenges as they require immediate response, which should be based on having trained and exercised on various emergency and business continuity plans (see Chapter 7 and the appendices).

In the U.S., under Occupational Safety and Health Administration (OSHA) regulation 1910.38(b), each workplace that has more than 10 employees must have a written emergency action plan (less than 10 employees may have a plan orally communicated to employees, for an example, see Appendix 1). Well planned, trained, and exercised emergency and business continuity plans that result in the implementation of critical function actions can mitigate negative impacts, lead directly to implementing continuity plans, reduce physical damage, and save lives (Xiao & Peacock, 2014). To consider how those plans relate to critical infrastructure, the following sections examine disruptions to specific systems and identify issues for planning and mitigation (covered further in Chapter 8 and Appendix 2). The present discussion will cover utilities (power, water and wastewater, natural gas), transportation systems and supply chains, and communications and information technology.

Utilities

Utilities include a range of services required by businesses to operate: power, water, wastewater, and natural gas to name the most critical. This section provides a brief overview of how utilities provide necessary resources to businesses, how disruptions may affect those businesses, and how some of them coped.

Power

As evidenced by the 2012 derecho example, electric power provides a foundational component of many other infrastructure sectors on which businesses rely. Power losses from the derecho led to cascading failures in communications, transportation, distribution components of the energy sector (oil and natural gas), public health, water, and emergency services, and government facilities. Such cascading failures have been seen since 2012 as well. For example, the fragility of the power grid in Puerto Rico in the 2017 hurricane season resulted in a recommendation that "governments at all levels and private sector owners of critical infrastructure need to further invest in resilient grids and prepare for outages" (DHS, 2018, p. iii).

The general architecture of the electric power grid includes connected components for power generation, transmission, and distribution, each supported by transformers that raise and lower voltage throughout the system (U.S. Department of Energy, 2015). Power disruptions may result from problems in any component of the power grid and could be localized to the facility site itself. Hazard vulnerability for the power grid ranges from natural hazards to cyber and physical attacks, as well as from operational error and component failure. However, severe weather causes most power outages in the United States (Executive Office of the President, 2013; National Academies of Sciences, Engineering, and Medicine, 2017). Power lines can also cause disasters, which has been the case with catastrophic wildfires in California and bushfires in Australia (Mitchell, 2013).

As a business owner or operator, you cannot control the offsite issues related to power availability, but you can plan and mitigate the impacts of power loss on critical functions. One vital action is to connect with the electric power provider, which will ensure that the company receives outage information and that your business knows the timeline for restoration. Knowing the duration of an expected outage helps to determine the scope of necessary business continuity actions. Based on the duration, actions could be handled as easily as temporary closure or canceling an incoming shift due to a short-term outage. For longer-term outages, businesses may need to transfer workloads to other operational sites or decide to cease specific functions or entire operations. Planning for a power outage should thus consider a variable time from hours to weeks when a business could be shut down. For example, hurricane Katrina shut down the power grid in New Orleans from weeks to over a year in various sections of the city. The most affected areas included the hospitals, where employees took heroic actions to save lives when power failures resulted, generators ran out of fuel or failed, and evacuations had to be undertaken in rapidly deteriorating conditions (see Fig. 5.1 and Box 5.3).

FIGURE 5.1
A military tanker is filled with diesel fuel for delivery to a hospital generator during the 2017 hurricane recovery in Puerto Rico.

BOX 5.3 New Orleans hospitals and Hurricane Katrina.

New Orleans hospitals faced impossible odds to save patients after hurricane Katrina left massive buildings stranded in the 2005 storm surge and flooding. Close to two dozen hospitals lost power, communications, water, wastewater, and natural gas utilities and had to evacuate patients (Gray & Hebert, 2007). Though some hospitals had generators, placement of the units in basements led to rapid failures and fuel supplies were used up within days. Patients on ventilators powered by the basement unit then received support from portable generators, including infants in one NICU. Requiring transfer to a more reliable site, a series of transportations ensued from a fire truck that could navigate some high water and ending with a canoe. The infants survived, thanks to innovative methods that occurred after multiple critical function failures (Barkemeyer, 2006). Tulane's hospitals also had generators, which ran out of fuel after 36 hours leading to temperatures more than 90 degrees. Staff members worked in shifts to hand-ventilate patients in respiratory distress (Taylor, 2007). At the 3-day point, a makeshift helicopter airlift began to evacuate patients from the parking lot roof. At Memorial hospital, 2500 patients had to be evacuated (many via helicopter) as temperatures reached over 100 degrees. Approximately 40 patients died at Memorial, later leading to investigations against some of the staff amid allegations of euthanasia (Fink, 2013).

To continue serving patients required widespread assistance. Federal medical stations, Medical Reserve volunteers, and Disaster Medical Assistance Teams (DMATs) arrived from outside the inundated area, setting up critical care centers for both patients and nursing home residents within affected and surrounding states. Mobile hospital units arrived from outside the state as well, providing diagnostic and outpatient services (Blackwell & Bosse, 2007). Medical volunteers also provided alternative sites for dialysis and cancer care as well as dental and optometric services. Katrina devastated an already weakened health care system and resulted in the permanent closure of New Orleans' hospital for the lowest income residents, Charity Hospital, demonstrating that impacts of infrastructure outages are not generally equitable (Coleman, Esmalian, & Mostafavi, 2020).

Resources, last accessed July 27, 2020: Barkemeyer, B. M. (2006). Practicing neonatology in a blackout: The University Hospital NICU in the midst of Hurricane Katrina: Caring for children without power or water. Pediatrics, 117(Suppl. 4), S369–S374; Blackwell, T., & Bosse, M. (2007). Use of an innovative design mobile hospital in the medical response to Hurricane Katrina. Annals of Emergency Medicine, 49(5), 580–588; Coleman, N., Esmalian, A., & Mostafavi, A. (2020). Equitable resilience in infrastructure systems: Empirical assessment of disparities in hardship experiences of vulnerable populations during service disruptions. Natural Hazards Review, 21(4), 04020034; Fink, S. (2013). Five days at memorial. NY: Penguin; Gray, B. H., & Hebert, K. (2007). Hospitals in Hurricane Katrina: Challenges facing custodial institutions in a disaster. Journal of Health Care for the Poor and Underserved, 18(2), 283–298; Rudowitz, R., Rowland, D., & Shartzer, A. (2006). Health care in New Orleans before and after Hurricane Katrina: The storm of 2005 exposed problems that had existed for years and made solutions more complex and difficult to obtain. Health Affairs, 25(Suppl. 1), W393–W406; Taylor, I. L. (2007). Hurricane Katrina's impact on Tulane's teaching hospitals. Transactions of the American Clinical and Climatological Association, 118, 69.

Facility plans should also outline immediate procedures for power loss from both a safety and continuity perspective. Preparations for power loss include installation and inspection of emergency lighting, protection of vital systems with surge protection, and uninterruptable power supplies. More costly mitigation efforts for electric power loss include the installation of on-site power generation capability, which can be reasonably cost-effective for a small business but significantly expensive (and essential) for a large hospital. Facility-level power generation may provide full or partial service targeted at critical systems such as computers or operating rooms. In either case, generators require regular maintenance and fuel availability, and that work teams be tasked, trained and competent in the specific steps necessary to both activate and deactivate

---

BOX 5.4 Power outage toolkit.

What can businesses do to prepare for a power outage? The process starts by asking questions about critical functions that rely on electricity, wind energy, or even solar power. For example, which business functions require power to operate such as the Internet, phone, computers, production lines, entrances to parking garages, and navigation systems? What about fueling vehicles or maintaining heating and cooling capabilities? Abilities to communicate may also be compromised.

By identifying which elements of the business rely on power, the first step has identified potential points of vulnerability (FEMA, 2015). The next question should ask if employees can continue to do their work during a power outage? As seen in this chapter, health care workers can still record patient information using pencil and paper and then enter that information when utility companies restore power. But, without a generator backup, they may not be able to restart a patient's heart using an AED or defibrillator. In addition, power outages may disrupt abilities to secure a facility without battery backups or emergency lighting. Essentially, how much of the BCP critical functions can be accomplished without power or a backup process in place to provide temporary power?

By focusing on critical functions, which keep revenue flowing into a business, planners can identify the high-risk failure points and design workarounds. Pre-placed power outage kits can be useful, and include flashlights and alternate power sources like batteries, lanterns, portable power chargers, and extra cell power banks. Payroll will need to continue and can operate via alternative ways to record hours, generate, and distribute pay. Solar power can be used to recharge cell phones as can hand-cranked batteries. Generators keep hospitals and nursing homes viable and can also power critical functions at restaurants, insurance companies, and hair salons. Generators should always be accompanied with a carbon monoxide detector to ensure safety. Computerized functions can remain online, though a business may need to narrow work activities down to the capacities of the alternative power sources or work without power. Alternative ways to communicate may be needed, in part to reduce reliance on alternative power sources needed for critical functions. Employees will need to be trained and, where critical functions require immediate action, to go through an exercise to test the training (see Chapter 7).

Resources, last accessed July 27, 2020: *FEMA (2015). Ready business power outage toolkit (https://www.fema.gov/media-library/assets/documents/152387); Red Cross (n.d.). Power outage safety. https://www.redcross.org/get-help/how-to-prepare-for-emergencies/types-of-emergencies/power-outage.html.*

---

systems. It is important that as facility use changes, such as the re-organization or relocation of work groups, production processes, and other workflow evolutions, businesses adjust power supply needs as well (see Box 5.4).

Water and wastewater

Water and wastewater disruptions can also cause catastrophic effects. Like electrical power, many water systems rely on a supply and deliver structure that may include reservoirs, water treatment plants, pumping stations, pipelines, and fire hydrants (Masoomi & van de Lindt, 2018). Depending on the water utility and system structures, water and wastewater systems often run parallel and are operated by the same utility. Some utilities may only provide water with wastewater handled through on-site septic systems that are periodically pumped and disposed of by licensed commercial carriers. Some systems may consist of both on-site well and septic systems. While loss of power may impact water and wastewater for prolonged outages, onsite well systems that are not supported by local power generation may become immediately unavailable with the loss of power. Like the power supply, individual businesses may rely on what local jurisdictions can produce and deliver, including the quantity and quality of the water. Disruptions may cause serious problems, especially for companies that rely on water and water quality. Depending on the nature of the business and/ or manufacturing processes it may be necessary for the facility to pre-treat water or post-treat water used in production processes before discharge back into systems (Brozović, Sundin, & Zilberman, 2007). Changes in water quality or source may cause larger problems for these businesses.

In 1995, the Kobe earthquake caused utility disruptions for months. Gas leaks led to multiple fires, which firefighters could not extinguish without a reliable water source. Hospitals lost critical function capabilities due to water loss that compromised both lifesaving and more routine health care delivery systems. Utility restoration took up to 3 months for this catastrophic earthquake that also destroyed bridges, highways, and railways (Schiff, 1999). Later, the 2011 tsunami and earthquake in Japan caused water disruptions in surrounding areas, with loss of the key utility for up to 20 days at hospitals outside the immediate impact areas which were destroyed. The disruption resulted in surviving hospitals being unable to sterilize equipment, offer services like dialysis, produce food for patients, and serve outpatients (Matsumura et al., 2015). Even more seriously, the Fukushima-Daiichi nuclear plant required water to maintain operations and avoid a catastrophic event. Conversely, plant failures led to widespread concerns and worldwide testing for contaminants to determine the viability of food and water supplies (Hamada, Ogino, & Fujimichi, 2012).

Mitigation options for water and wastewater failures can be costly. These differ radically based upon the nature of the business. Beyond the storage of bottled drinking water on-site, the addition and maintenance of bulk on-site storage can be costly and if not maintained to public health standards, can result in negative health effects. For wastewater, options include temporary onsite storage solutions for water used in manufacturing processes. For sanitary wastewater, options include maintaining contracts for emergency portable toilets and/ or restroom trailers.

Natural gas and other fuels

Natural gas flows in a pattern like other utilities. Before it reaches a business, natural gas flows from its source through a series of interstate and intrastate pipelines (see naturalgas.org). Gas can then be stored or moved through pipelines via compressure, metering, and control stations. With the gas off, business owners reduce risks but then face economic impacts when some critical functions become inoperable. One study found that business owners thought they could get by for about a week without natural gas and ranked electricity, communications, and water higher (Nigg, 1995). Areas that rely on natural gas for heating, though, would find that a loss of this utility could undermine life safety in a polar vortex. The hazards identification planners accomplished in Chapter 3 should have revealed potential situations in which loss of this utility would prove problematic.

While one of the most reliable utilities, natural gas also represents a potential threat in some hazards. Earthquakes, for example, require that homeowners and business owners know how to turn off the gas to avoid fires and explosions. The 2018 Merrimack Valley natural gas system failure caused explosions across three towns in Massachusetts. The resulting recovery including businesses that were directly affected when post-incident reconstruction of gas lines resulted in one-third of businesses not operating like they were before the disaster (Walters, 2019). Fuels necessary for business processes can include heating oil and other fuels that required to maintain a vehicle fleet. In 2012, Hurricane Sandy disrupted supply networks across states in the U.S. Northeast when flooding and high winds caused power interruptions that affected distribution networks. Both refineries and pipelines sustained damage, with downstream effects for fuel delivery and consumption that affected business viability. Overall, Sandy became one of the most economically dangerous storms in U.S. history, causing approximately $100 billion in damage with flooding to thousands of buildings and businesses (Haraguchi & Kim, 2016).

Communications and information technology

For smaller businesses and some larger companies, communications and information technology functions often originate with the same provider, such as a company that provides Internet and Voice of Internet Protocol (VoIP). In such single provider systems, an Internet connectivity outage also disrupts voice communication systems as these can be tightly coupled and interdependent systems, thus subject to more significant impacts. Communications and Information Technology components also share similar vulnerabilities like electric power, as many wired systems share the same space on poles and towers. Voice, video, and data all transmit through the core network to access points for broadcasting, cable, satellite, wireless, and wireline. Communications disruptions can be difficult based upon the nature of the outage, characteristics of the communication network at the facility, and the business needs.

Hurricane Katrina's high winds and flooding compromised communications for some time, including cellular and landline services needed for live-saving efforts. Both emergency and calls for family wellness checks could not be accomplished and those who did try to use such services often overwhelmed available services. Most cellular service companies sustained damage, with some operating through limited generator availability and then through temporary cell sites on wheels. Texting worked first if a signal could be acquired. Matters have improved since Katrina, though superstorm Sandy also undermined communications capabilities in part due to power outages (Kwasinski, 2013). By Sandy in 2012, communications had shifted significantly with an increase in app usage including to transmit warnings, identify shelter locations, and share information via social media platforms and Internet-based websites. While the cellular system remained more robust than with Katrina, power outages reduced abilities of users for a variety of business and personal purposes. The 2011 Japanese earthquake and tsunami demonstrated the increasing interconnectivity of critical infrastructure even further. Cellular communications were disrupted for months and battery backups failed due to the extended outage. Nearly 5000 mobile base stations were impacted. A massive effort resulted, using over 11,000 workers to repair most of the damage, an effort made more difficult by damaged transportation arteries (Kobayashi, 2014).

Moving forward in time, the 2020 pandemic revealed even greater interconnectedness between critical infrastructure systems, particularly communications and information technologies. Massive numbers of businesses, agencies, organizations, schools, and universities increased their reliance on communications and IT as a survival measure. Alarmingly, cyberattacks and cybercrimes increased as well, attempting to penetrate shallower defenses with people working from home. The word "zoom-booming" became a phrase where intruders penetrated video platforms to insert inappropriate images, eavesdrop, or to take over a meeting. Video platforms put additional security measures into place, including passwords and required hosts to admit participants (Mohanty & Yaqub, 2020). Increases were observed in phishing, where cybercriminals victimize online users to secure personal and business information followed by identity theft and penetration of formerly secure systems (Ahmad, 2020).

To prevent disruptions, put backup systems into place. For example, VOIP phone systems can be supported by traditional wired phone lines, cellular, or satellite backup systems though they can be costly. Some disasters disrupt cellular service, but texting availability remains. Even older forms of communication can suffice including in-person meetings, use of whiteboards, and notes as well as walkie-talkies and ham radio operator services. New Orleans-area businesses that reopened after Katrina used signboards set out in front of their enterprises because the Internet system had not yet been restored. For Japan, companies put temporary public phones into various locations and gave access to tablets in affected areas (Kobayashi, 2014). The 2010 Haiti earthquake saw the installation of temporary charging stations in relief camps that were set up for years.

Businesses and agencies can also step up their internal educational processes about cybercrimes, from phishing to meeting intrusions. Employees should be advised to backup their computerized and server information on alternate sites regularly, preferably daily. Offsite services can be secured to offer such backups. Employees should also be instructed to change their passwords frequently and to keep them in a secure location, preferably one that is not written down though some apps do provide for password storage. Software should be kept updated regularly to eliminate errors that allow for intrusions. Virus software should not only be installed but updated regularly and monitored for threats. Businesses should be alert to hackers and attacks that occur worldwide to stave off a possible localized crime. Multi-factor authentication, which requires multiple steps to log in to an account, should be introduced as an additional barrier.

For broader emergency situations that may impact the availability of communications systems, the federal government maintains the Government Emergency Telecommunications Service (GETS) that provides access to special systems to ensure connectivity. This program is available to U.S. business that provide core infrastructure services and extends to cellular networks for priority service under the Wireless Priority Service (WPS) program.

Transportation systems and supply chains

As noted in the examples from previous chapters, transportation arteries represent crucial supply chain assets for businesses. Transportation services move goods and services, deliver needed resources, and enable business travel. High-level dependency on trucking, trains, planes, and ships means that any disaster might disrupt such deliveries while also impacting the abilities of employees to get to and from work.

For example, in 2002, a bridge collapsed on Interstate 40 in Oklahoma in the U.S. Freight flow disruptions affected business operations across all 50 states, with the most impact on nearby regions. East to west transportation flows changed as trucking companies sought alternative arteries to deliver goods. Consequences also included increasing congestion elsewhere as well as raising a company's fuel costs and delivery times which get passed on to those contracting with delivery services (Aydin, Shen, & Pulat, 2012). Such transportation failures also compromise emergency services and commuting, by lengthening response time required for first responders to reach someone who is injured or ill (Chang, 2016).

Businesses that rely on supply chains operating through transportation systems need backup plans. Earthquakes disrupt supply chains, as they buckle roadways, railways, and ports. Thus, even an earthquake distant from a specific business may cause problems with needed materials. For example, the 2011 Japanese tsunami resulted in $210 billion in supply chain disruptions while the 2011 New Zealand earthquake caused $20 billion worldwide. Terrorism also disrupts supply chains, as efforts to avoid seaborne pirates have caused an estimated $642 million in losses from avoidance of the Suez Canal (Kungwalsong & Ravindran, 2012). Businesses must act to protect the production of goods and services and abilities to conduct business travel as part of their business continuity planning.

Most recently, the coronavirus slowed interstate delivery systems in many countries, a situation that continued for some time as various sites within affected nations established travel restrictions and mandated quarantines. Potato farmers in the U.S., for example, saw their produce rot while unable to deliver to empty grocery store shelves. Cleaning supplies then fell into short supply, a situation that worsened as nations and affected areas reopened their enterprises. Essential service sites, like grocery stores, may have benefited the most by being open as they were able to purchase critical items like Plexiglas to protect employees. Universities and schools that opened later in the year saw shortages and had to triage which area received installations. In terms of the overall impacts, such disruptions affect single businesses and the world economy due to the globalization of supply chains when needed materials are secured from outside one nation – only to address shortages there because of the worldwide impact of the pandemic.

Strategies to handle supply chain issues include having an on-hand and updated inventory of resources needed for critical functions. That inventory should be monitored daily so that adjustments can be made including expected downtime and employee workloads. The supply chain can also be monitored to varying degrees to make strategic choices about critical functions. Sector network mapping, which can be time-consuming, identifies a key product to the lead supplier, their suppliers, and then to producers of raw materials (Choi, Rogers, & Vakil, 2020). By doing such mapping, a business can identify the various vendor points where disruptions might occur. Business continuity planning directs planning teams to then identify the weakness of contracted vendors but also the ways in which those vendors expect to recover from a disaster. Backup vendors should be identified to maintain the flow of operations. Businesses should also monitor the supply chain daily to anticipate the ebbs and flows of supplies. COVID-19 made such monitoring an important task for health care providers attempting to secure personal protective equipment and cleaning supplies. Such monitoring, while tedious, can reduce the time that businesses will spend offline and can lead to appropriate adaptations during a crisis.

Downtime management

Downtime, when businesses cannot operate due to critical infrastructure outages or shortages, can be costly. Recent estimates indicate that thousands of businesses failed during COVID-19. At the start of the event, one national study in the U.S. found that the median business had $10,000 in monthly expenses but only 2 weeks of cash available to offset downtime losses, survive closures, and pay employees (Bartik et al., 2020). Businesses attempted to address losses from forced closures by converting to online options, a choice that could prove problematic as noted earlier. But even online orders dropped for some businesses, though other online services escalated dramatically, especially online food ordering (Chang & Meyerhoefer, 2020; Hasanat et al., 2020). Managing downtime is likely to require a significant amount of flexibility to become resilient (Graveline & Gremont, 2017).

Downtime related to critical infrastructure can vary greatly depending on the magnitude of an event, the structural and nonstructural damage that occurs, and the status of the utilities in each area (De Iuliis, Kammouh, Cimellaro, & Tesfamariam, 2019; Kammouh, Cimellaro, & Mahin, 2018). As noted in the American Civil Society of Engineers' most recent report card on the United States, critical infrastructure remains exceedingly fragile in many areas (ASCE, 2017), earning an overall grade of D+. Infrastructure grades of interest to business owners include the transportation sector (bridges C+, rail B, roads D), protective structures (dams and levees, D), and utilities (water D, wastewater D+, energy D+). Clearly, businesses will need to be prepared to deal with a range of potential problems and outages affecting both on-site production and delivery of goods and services as well as receiving resources through a dispersed supply chain subject to disruptions anywhere.

How utilities get back on

Getting utilities back online typically starts with reporting the outage to the company providing such services. The process occurs in overlapping steps (National Academies, 2017, p. 110):

- Assess the extent, locations, and severity of damage to the electricity system;
- Provide the physical and human resources required for repairs;
- Prioritize sites/components for repair based on factors including the criticality of the load and the availability of resources to complete the needed repairs; and
- Implement the needed repairs and reassess system state.

The priority of restoration varies but typically commences with critical infrastructure including hospitals, first responder units (police, fire, emergency medical, emergency managers), water pumping stations, and communications. Once these sites have been restored, companies will turn to (or simultaneously assign teams to) neighborhoods and businesses that have been affected. Some utilities operate from the most dense areas to the least depending on the outage.

Utility outages, if sufficiently large and widespread, will likely result in help arriving from other areas. After every major hurricane in the U.S., utility companies sent teams of paid and volunteer workers to affected areas. Military units have also arrived to assist with generators, fuel, communications capabilities, and water. Typically, utility workers will swarm into an area, reconnecting and repairing downed and disrupted lines in an areal pattern. They will also work to restore high-priority locations like hospitals, nursing homes, and similar areas requiring life-saving utilities.

The role of critical function work teams

Downtime from utilities can vary significantly, from a few hours after a small windstorm to months after a major disaster. Businesses will need to be aware of area hazards and the range of outage impacts that can occur to anticipate and repel the most serious of consequences. Whether a business needs to turn on a generator to save the food at a restaurant to ensuring that life-saving devices can continue to be powered, utility restoration and downtime management represents a very important critical function. Work teams should be assigned specifically to utility restoration and management whether that is a single employee tasked with reporting and monitoring the outage to turning on and fueling a small generator to entire shifts dedicated to maintaining power for a large hospital.

Critical function work teams, coupled with appropriate training, can make a difference. For example, many sites rely on electronic recording of records for customers, patients, and students. Power and/or Internet outages threaten recording such vital information that can be life-saving such as patient allergies to specific drugs. Workarounds tend to be old school, but they do enable patient care including using face-to-face communication, writing out records by hand, and building teams with go-kits of hard copy records. In 2017, an Australian hospital did that by using communication strategies that traveled through critical function work teams (e.g., nursing and pharmacy) and by using the vertical organization of the health care setting (Dave, Boorman, & Walker, 2020).

It is also likely that a business has already experienced downtime, such as loss of power, an Internet disruption, or a production line that stopped unexpectedly. Planners should therefore look internally to see how such downtime was handled previously, so that workarounds that surfaced can be reviewed for their value in a larger disaster. This is the moment where the diversity of a planning team matters, because long-time employees may be the ones who have the best workarounds. Given our increasing reliance on technology, and the interconnectivity of critical infrastructure systems to support that reliance, a wide set of planners will be needed to surface what worked before that dependency occurred.

Displacement management

In reviewing the research literature and considering our own observations from several decades in emergency management, several types of displacement may occur. Those types include site, function, and population displacement.

Site displacement

Site displacement happens when a physical plant, building, or rented space becomes unavailable. When the 2004 tsunami flooded the Naggapatinam, India, hospital compound, water entered close to 60 buildings. The water rose over the height of patient beds and inundated the intensive care unit. Despite the sudden onset event, staff and family members carried patients to second floors and no one perished. The hospital then mucked out the damaged rooms, re-established same-day care, and continued functioning under difficult conditions. To reduce future risks, they moved critical care units further inland. It was an impressive effort in a short period of time and in a place with scarce resources. It worked because the hospital team had previously trained for an emergency not knowing it would happen that day in such a traumatic fashion. Though the hospital did reopen the same day, they also moved functions further inland over the next 6 months to prevent a future disaster.

Options for managing site displacement include moving operations to a completely new location. The Joplin, Missouri St. John's Regional Medical Center did this when the EF5 tornado heavily damaged their nine-story building structure, ended any chance of using utilities for the foreseeable future, cut off communications and internet services, and downed trees across needed transportation arteries and power lines. Hospital employees and volunteers evacuated patients and managed a patient surge among 1000+ injured by the tornado. In addition to evacuating patients to another hospital, staff activated their disaster plan and set up a temporary emergency department at a city building. A mobile army surgical hospital unit then arrived to provide field-based support for those who had been injured or had become ill after the tornado (Ruth, 2011). Pre-planning is what enabled the medical professionals to save lives and continue their work, including transitioning to new and temporary locations within hours.

Other options might be moving the critical function to another location within the physical plant of the business, which is what some New Orleans' hospitals did during hurricane Katrina. The 2016 Kumamoto, Japan earthquakes caused building damage to multiple industries including the automotive and chemical sectors. By working to rapidly restore utilities and using alternative sites, businesses salvaged customer relations successfully (Maruya & Torayashiki, 2017). Site displacement may also happen in unusual ways, such as when faith traditions lost their physical site for offering worship services during the COVID-19 pandemic (see Box 5.5). While pre-planning for site displacement works best, faith traditions demonstrated flexibility and innovation that enabled them to transition.

Functional displacement

Functional displacement happens when a specific critical function cannot continue because of the disaster. To deal with functional displacement, businesses may opt to outsource, subcontract, downsize, temporarily cease, or stop specific functions. For example, a dairy operation might lose a specific barn that produces yogurt but still have other production capabilities. While the barn is being replaced, production of yogurt might be discontinued while milk and other products can still be created. Or, a dairy operation might rely on a competitor to continue yogurt production with a subcontract. When disasters happen, similar businesses also often step up and provide resources and space. Government agencies may also provide space to use in an emergency so that businesses can continue.

Regardless, potential functional displacements must be considered in the business continuity plan. A pandemic, for example, could undermine an entire unit like payroll which the company needs to survive. Backup employees could be brought in from a temporary agency, a similar business, or preferably another unit of the business situated in another location. Cross-training for the critical function of payroll will have been crucial (see Chapter 7) as will the ability to conduct such operations remotely. A business may also need to consider transferring operations elsewhere like when hospitals or nursing homes relocate patients to alternate facilities. When hurricanes approach the U.S. coast, some nursing homes move patients further inland to alternate sites less likely to lose critical infrastructure. Such movement, though, has to be carefully considered because the possibility of transfer trauma, which occurs when fragile and highly vulnerable patients cannot withstand such movement (Fernandez, Byard, Lin, Benson, & Barbera, 2002). This kind of movement requires a significant investment of pre-planning, time, resources, and people.

Population displacement

Population displacement typically refers to a loss of workers due to a disaster. A loss of human resources occurred after hurricane Katrina, when workers became displaced across the United States. Employers struggled to find qualified workers able to work under difficult conditions, because employees could not return due to damaged transportation systems and the loss of their homes. Employers also competed with others, including the U.S. government, to hire people at affordable wages. While human resources will be discussed further in

BOX 5.5 Faith traditions and displacement.

Among the businesses, agencies, and organizations displaced by COVID-19 were the followers of the world's major faith traditions. Faith traditions include multiple critical functions providing social solidarity, social control, and answers to life's questions (Durkheim, 1915). Religions meet these critical functions through an array of rituals and beliefs acted out specific to a faith tradition or denomination. COVID-19 disrupted those critical functions with a heavy impact to the place-based rituals essential to many religions. Adaptations required critical infrastructure including utilities, communications, and Internet connectivity as faith traditions moved onto alternative platforms.

Stay-at-home restrictions led to closures of mosques, synagogues, churches, temples, and places of worship worldwide. Centralized worship resources came from organizations serving faith traditions including alternatives for online sermons with videos of people singing hymns, and readings of scripture that could be imported into more locally created video productions. For followers of Islam, the five-time daily call to prayers included messages to observe daily obligations at home. Live-streaming, microphones, and online services provided access to the faithful and not only for Islam. Christians around the world turned to services provided over social media, video platforms, and livestreaming. As the pandemic continued, more innovative options surfaced including parking lot car worship, drive-in theaters, and televised delivery. Traditional services like Judaism's Erev Shabbat and Torah study moved onto the less convention use of Zoom. Funeral traditions shifted as well. Buddhists, who bury or cremate within 24 hours, turned to services that protected the living from any virus in the deceased. Congregational conferences opted into virtual platforms for their annual and committee work. World religions cancelled historic, annual pilgrimages for Hindu and Muslims while the ecumenical calendar for Christians moved regular communion services and events like Easter into virtual events conducted on people's dinner tables at home.

In traditions where personal contact and social connectedness represent valued and normative behaviors, the virus challenged displacement and how to remain viable in new ways to interact. With a physical setting lost for an extended time being, faith leaders looked for new ways to engage in meaningful discourse. Outreach efforts required new distancing, such as ministering to the homebound or nursing-home bound faithful through windows or video interactions. Spiritual ministering moved online. Members recorded personal messages for placement on social media platforms to be inserted into worship services. Fellowship opportunities shifted onto Google hangout, Zoom, and similar ways to remain connected. As the pandemic lingered, worship centers purchased low-cost devices to enable participation among those who had become increasingly socially isolated. Volunteers acting on their spiritual convictions expanded further into helping those who lost jobs, homes, and food sources.

Ultimately, faith traditions came to view the displacement as a challenge to be embraced, by protecting the highly vulnerable among their followers including both paid and volunteer staff. Reopening requirements varied from newly introduced social distancing to choirs wearing masks or faith leaders delivering sermons via loudspeakers and other alternative means. Rather than a handshake or a hug, traditions evolved into nodding and holding hands in a prayerful display. Given that all faith traditions address disaster and tragedy, using scriptures and belief systems ultimately relied on their traditional theology to meet the critical functions for those experiencing displacement.

Resources used for this figure, all last accessed July 26, 2020: *Durkheim, E. (1915). Elementary forms of religious life. London: George Allen & Unwin; Ibrahim, A. (2020).* Praying in time of COVID-19: How world's largest mosques adapted. *Available at https://www.aljazeera.com/amp/news/2020/04/praying-time-covid-19-world-largest-mosques-adapted-200406112601868.html; ICRC (2020).* Buddhist management of the dead and COVID 19. *https://reliefweb.int/sites/reliefweb.int/files/resources/002_buddhist_management_of_the_dead_and_covid-19_web.pdf; Sekhsaria, K. (2020).* How Hindu spiritual practices can help manage your COVID-19 anxiety. *Available at https://www.hinduamerican.org/blog/hindu-spiritual-practices-manage-anxiety-covid; Temple Beth-El (2020).* Coronavirus COVID-19. *Available at https://tbe-sb.org/; United Church of Christ (2020).* Worship resources. *Available at https://docs.google.com/document/d/1E_RKfy_VOzelb5jbXV7UW5ytjmil-_Md-W0H8UaRIaM/edit?ts=5e66621c [for similar documents, search the name of the faith tradition and COVID-19].*

Chapter 6, some strategies for addressing population displacement include hosting job fairs to find employees, hiring temporary workers, bringing back retired employees, and identifying backup workers as part of a business continuity plan. Outside sources of support can also aid a business, such as when health care workers went into COVID hot zones to support exhausted staff dealing with increasing hospitalization rates. Businesses may also need to opt for reduced or cancelled shifts, offer flextime, outsource, or reduce the scope of a workload due to employee unavailability. Employees, and their backups, represent the most important resource necessary to getting things back toward normal.

Restoring critical functions

Business continuity plans include identifying critical functions at the heart of the planning effort. When writing critical functions, care must be taken to consider the kinds of disruptions that may occur with critical infrastructure. For example, how would loss of power affect the critical functions identified by planners. Can payroll still be done? Will cows still be milked at a dairy operation? Can a dry cleaner continue operating? When will food in a restaurant freezer become compromised? How long can the business endure such a power outage without serious damage to daily, weekly, or monthly revenue? What options exist, such as a generator, to continue operations or salvage resources?

Writing critical functions with critical infrastructure in mind requires asking and answering questions, such as:

- What will be disrupted if we have a power outage?
- What will be disrupted if we have a water, wastewater, or natural gas shortage?
- How will a business deliver its goods and services if its primary transportation arteries change?
- What is the breaking point for a cost-benefit on fuel consumption or the availability of drivers?
- Will additional time spent in delivering items (e.g., refrigerated food trucks) compromise the quality of the food or the nature of the contract?
- For critical deliveries, could air carriers be used in place of trucking operations or vice versa?
- How will the supply chain to the business be affected?
- Which operations might be affected due to supply chain disruptions?
- Who is the alternate vendor and how long will it take to secure those resources?

- Will some critical functions need to be temporarily reduced or stopped due to shortages? If so, at what point in time can the business tolerate that disruption?
- What happens if an Internet disruption occurs? What workarounds can be used from alternate technology to non-tech solutions?
- How will we communicate if landline and cellular phones are out?
- Will additional personnel be needed, for example, to register students at a university if registration systems fall to a cyberattack?
- How will a hospital (or similar enterprise) record patient records, order pharmacy supplies, or request meal delivery to patients?
- Where should specific functions be moved to if disrupted? What resources and people would it take to make that relocation happen effectively?
- Where are some alternate sites to continue operations if a building becomes permanently compromised?
- What happens if our workforce is out due to injuries or illness? Who needs to be cross trained in areas that rely on critical infrastructure? Who are their backups?
- What are the workarounds to continue with essential operations and for how long?
- What is the breaking point at which operations simply need to be shut down and for how long?

Once such questions have been considered specific to the business, work teams should be assigned and trained for the critical function. A critical function that relies on power should have a team that contacts the utility to determine restoration time, communicates that timeframe to the workers in the unit, and/or establishes backup power through turning on and maintaining a generator.

Essential actions

- Establish clear lines of communications with utilities on which your business relies and learn their protocol to report disruptions and how they plan to restore functionality.
- Consider backup power sources such as a generator, batteries, solar power and other ways to continue critical functions.
- Identify multiple backup procedures for communications and connectivity.
- Identify alternate transportation routes and options for supply chain into the business as well as for delivery of goods and services both physically and electronically.
- Set up work teams for specific critical functions and train those teams.
- Identify backup workers for critical functions and provide cross-training.

- List out vendors and contact information in the plan and include in a go-kit if an evacuation is needed. Identify backup vendors should the supply chain become compromised.
- Pre-locate possible sites for both entire industry and function-specific displacement. Develop agreements with those sites should relocation become necessary. Outline protocol for movement of industry and/or functions to the temporary site.
- Pre-identify sources of temporary workers to address population displacement that affects operating capabilities.

References

Ahmad, T. (2020). *Corona virus (COVID-19) pandemic and work from home: Challenges of cybercrimes and cybersecurity.* Available at SSRN 3568830.

ASCE (2017). *Infrastructure report card.* Available at *(2017)*. https://www.infrastructurereportcard.org/. (Last Accessed July 26, 2020).

Aydin, S. G., Shen, G., & Pulat, P. (2012). A retro-analysis of I-40 bridge collapse on freight movement in the US highway network using GIS and assignment models. *International Journal of Transportation Science and Technology, 1*(4), 379–397.

Barkemeyer, B. M. (2006). Practicing neonatology in a blackout: The University Hospital NICU in the midst of Hurricane Katrina: Caring for children without power or water. *Pediatrics, 117* (Suppl. 4), S369–S374.

Bartik, A. W., Bertrand, M., Cullen, Z., Glaeser, E. L., Luca, M., & Stanton, C. (2020). The impact of COVID-19 on small business outcomes and expectations. *Proceedings of the National Academy of Sciences, 117*(30), 17656–17666.

Blackwell, T., & Bosse, M. (2007). Use of an innovative design mobile hospital in the medical response to Hurricane Katrina. *Annals of Emergency Medicine, 49*(5), 580–588.

Brozović, N., Sundin, D., & Zilberman, D. (2007). Estimating business and residential water supply interruption losses from catastrophic events. *Water Resources Research, 43*(8). Retrieved 27 July from(2007). https://agupubs.onlinelibrary.wiley.com/doi/10.1029/2005WR004782.

Chang, H. H., & Meyerhoefer, C. (2020). *COVID-19 and the demand for online food shopping services: Empirical evidence from Taiwan.* (No. w27427) National Bureau of Economic Research.

Chang, S. E. (2016). Socioeconomic impacts of infrastructure disruptions. *Oxford Research Encyclopedia of Natural Hazard Science,*. Retrieved 16 June 2020, from (2016). https://oxfordre.com/naturalhazardscience/view/10.1093/acrefore/9780199389407.001.0001/acrefore-9780199389407-e-66.

Choi, T., Rogers, D., & Vakil, B. (2020). *Coronavirus is a wake-up call for supply chain management. Available at (2020).* https://hbr.org/2020/03/coronavirus-is-a-wake-up-call-for-supply-chain-management. (Last Accessed July 28, 2020).

Coleman, N., Esmalian, A., & Mostafavi, A. (2020). Equitable resilience in infrastructure systems: Empirical assessment of disparities in hardship experiences of vulnerable populations during service disruptions. *Natural Hazards Review. 21*(4). https://doi.org/10.1061/(ASCE)NH.1527-6996.0000401.

Dave, K., Boorman, R. J., & Walker, R. M. (2020). Management of a critical downtime event involving integrated electronic health record. *Collegian.* https://doi.org/10.1016/j.colegn.2020.02.002.

De Iuliis, M., Kammouh, O., Cimellaro, G. P., & Tesfamariam, S. (2019). Resilience of the built environment: A methodology to estimate the downtime of building structures using fuzzy logic. In *Resilient structures and infrastructure* (pp. 47–76). Singapore: Springer.

Durkheim, E. (1915). *Elementary forms of religious life.* London: George Allen & Unwin.

Executive Office of the President (2013). *Economic benefits of increasing electric grid resilience to weather outages. Retrieved from (2013).* https://www.energy.gov/sites/prod/files/2013/08/f2/Grid%20Resiliency%20Report_FINAL.pdf.

Faupel, C. E., Kelley, S. P., & Petee, T. (1992). The impact of disaster education on household preparedness for Hurricane Hugo. *International Journal of Mass Emergencies and Disasters, 10*(5), 5–24.

Federal Emergency Management Agency (2019). *Community Lifelines Implementation Toolkit: Comprehensive information and resources for implementing lifelines during incident response. Retrieved from (2019).* https://www.fema.gov/media-library-data/1576770152678-87196e4c3d091f0319da967cf47ffd9c/CommunityLifelinesToolkit2.0v2.pdf.

FEMA (2015). *Ready business power outage toolkit. (2015).* https://www.fema.gov/media-library/assets/documents/152387.

Fernandez, L. S., Byard, D., Lin, C. C., Benson, S., & Barbera, J. A. (2002). Frail elderly as disaster victims: Emergency management strategies. *Prehospital and Disaster Medicine, 17*(2), 67–74.

Fink, S. (2013). *Five days at memorial.* NY: Penguin.

Goodrich, J. N. (2005). The big American blackout of 2003: A record of the events and impacts on USA travel and tourism. *Journal of Travel & Tourism Marketing, 18*(2), 31–37.

Graveline, N., & Gremont, M. (2017). Measuring and understanding the microeconomic resilience of businesses to lifeline service interruptions due to natural disasters. *International Journal of Disaster Risk Reduction, 24*, 526–538.

Gray, B. H., & Hebert, K. (2007). Hospitals in hurricane Katrina: Challenges facing custodial institutions in a disaster. *Journal of Health Care for the Poor and Underserved, 18*(2), 283–298.

Hamada, N., Ogino, H., & Fujimichi, Y. (2012). Safety regulations of food and water implemented in the first year following the Fukushima nuclear accident. *Journal of Radiation Research, 53*(5), 641–671.

Haraguchi, M., & Kim, S. (2016). Critical infrastructure interdependence in New York City during Hurricane Sandy. *International Journal of Disaster Resilience in the Built Environment, 7*(2), 133–143.

Hasanat, M. W., Hoque, A., Shikha, F. A., Anwar, M., Hamid, A. B. A., & Tat, H. H. (2020). The impact of coronavirus (COVID-19) on E-business in Malaysia. *Asian Journal of Multidisciplinary Studies, 3*(1), 85–90.

Kammouh, O., Cimellaro, G. P., & Mahin, S. A. (2018). Downtime estimation and analysis of lifelines after an earthquake. *Engineering Structures, 173*, 393–403.

Kobayashi, M. (2014). Experience of infrastructure damage caused by the Great East Japan Earthquake and countermeasures against future disasters. *IEEE Communications Magazine, 52*(3), 23–29.

Kungwalsong, K., & Ravindran, A. R. (2012). Modeling supply chain disruption risk management. In *IIE annual conference. Proceedings* (p. 1). Institute of Industrial and Systems Engineers (IISE).

Kwasinski, A. (October 2013). Effects of hurricanes Isaac and Sandy on data and communications power infrastructure. In: *Intelec 2013; 35th international telecommunications energy conference, smart power and efficiency,* VDE, pp. 1–6.

Lin, S., Fletcher, B. A., Luo, M., Chinery, R., & Hwang, S. A. (2011). Health impact in New York City during the Northeastern blackout of 2003. *Public Health Reports, 126*(3), 384–393.

Maruya, H., & Torayashiki, T. (2017). Damage of enterprises and their business continuity in the 2016 Kumamoto earthquake. *Journal of Disaster Research, 12*(sp), 688–695.

Masoomi, H., & van de Lindt, J. W. (2018). Restoration and functionality assessment of a community subjected to tornado hazard. *Structure and Infrastructure Engineering, 14*(3), 275–291.

Matsumura, T., Osaki, S., Kudo, D., Furukawa, H., Nakagawa, A., Abe, Y., … Kushimoto, S. (2015). Water supply facility damage and water resource operation at disaster base hospitals in Miyagi prefecture in the wake of the great East Japan Earthquake. *Prehospital and Disaster Medicine, 30* (2), 193.

Mitchell, J. W. (2013). Power line failures and catastrophic wildfires under extreme weather conditions. *Engineering Failure Analysis, 35*, 726–735.

Mohanty, M., & Yaqub, W. (2020). *Towards seamless authentication for Zoom-based online teaching and meeting*. arXiv preprint arXiv:2005.10553.

National Academies of Sciences, Engineering, and Medicine (2017). *Enhancing the resilience of the nation's electricity system*. Washington, DC: The National Academies Press https://doi.org/10.17226/24836.

National Infrastructure Advisory Council (2010). *A framework for establishing critical infrastructure resilience goals. Available at(2010).* https://www.cisa.gov/sites/default/files/publications/niac-framework-establishing-resilience-goals-final-report-10-19-10-508.pdf. (Last Accessed July 26, 3030).

National Infrastructure Advisory Council (2013). *Strengthening regional resilience: Final report and recommendations. Retrieved from(2013).* https://www.cisa.gov/sites/default/files/publications/niac-regional-resilience-final-report-11-21-13-508.pdf.

New Zealand Lifelines Council (2020). *New Zealand critical lifelines infrastructure national vulnerability assessment summary report. Available at (2020).* https://www.civildefence.govt.nz/assets/Uploads/lifelines/nzlc-nva-2020-summary.pdf. (Last Accessed July 27, 2020).

Nigg, J. M. (1995). *Business disruption due to earthquake-induced lifeline interruption.* (Preliminary Paper #220) Newark, DE: Disaster Research Center, University of Delaware.

Obama, B. H. (2013). *Presidential policy directive 21: Critical infrastructure security and resilience. (2013).* https://obamawhitehouse.archives.gov/the-press-office/2013/02/12/presidential-policy-directive-critical-infrastructure-security-and-resil.

Rose, A., Oladosu, G., & Liao, S. Y. (2007). Business interruption impacts of a terrorist attack on the electric power system of Los Angeles: Customer resilience to a total blackout. *Risk Analysis: An International Journal, 27*(3), 513–531.

Ruth, S. (2011). *Tornado pummels Joplin hospital. Available at (2011). https://journals.lww.com/em-news/FullText/2011/10000/Breaking_News__Tornado_Pummels_Joplin_Hospital,.4.aspx. (Last Accessed July 28, 2020).*

Schiff, A. J. (1999). *Hyogoken-Nanbu (Kobe) earthquake of January 17, 1995: Lifeline performance.* ASCE.

Taylor, I. L. (2007). Hurricane Katrina's impact on Tulane's teaching hospitals. *Transactions of the American Clinical and Climatological Association, 118*, 69.

Tierney, K. J. (1997). Business impacts of the Northridge earthquake. *Journal of Contingencies and Crisis Management, 5*(2), 87–97.

U.S. Department of Commerce (2013). *Service assessment: The historic derecho of June 29, 2012. Retrieved from(2013).* https://www.weather.gov/media/publications/assessments/derecho12.pdf.

U.S. Department of Commerce (2018). *About derechos. Retrieved from (2018).* https://www.spc.noaa.gov/misc/AbtDerechos/derechofacts.htm.

U.S. Department of Energy (2015). *Electricity industry primer. Retrieved from (2015).* https://www.energy.gov/sites/prod/files/2015/12/f28/united-states-electricity-industry-primer.pdf.

U.S. Department of Homeland Security (2018). *2017 Hurricane season after-action report. Retrieved from (2018).* https://www.fema.gov/media-library-data/1531743865541-d16794d43d3082544435 e1471da07880/2017FEMAHurricaneAAR.pdf.

U.S. Department of Homeland Security (2019b). *National response framework* (4th ed.). Retrieved from (2019b). https://www.fema.gov/media-library-data/1582825590194-2f000855d442 fc3c9f18547d1468990d/NRF_FINALApproved_508_2011028v1040.pdf.

U.S. Department of Homeland Security (2019c). *A guide to critical infrastructure security and resilience. Retrieved from (2019c).* https://www.cisa.gov/sites/default/files/publications/Guide-Critical-Infrastructure-Security-Resilience-110819-508v2.pdf.

Walters, Q. (July 9, 2019). *Businesses still struggle to bounce back from Columbia gas explosions after nearly a year. Retrieved from (July 9, 2019).* https://www.wbur.org/news/2019/07/09/columbia-gas-explosions-business-struggles-rock-the-register.

Xiao, Y., & Peacock, W. (2014). Do hazard mitigation and preparedness reduce physical damage to businesses in disasters? Critical role of business disaster planning. *Natural Hazards Review, 15,* 04014007.

CHAPTER 6

Managing human resources

Introduction

In this chapter, we look at the most valuable part of any business: the employees, the talents they use to support a business, and the importance of the human resource management team. Whether that employee stocks grocery store shelves in a pandemic or serves as CEO of a major hospital, both provide value to what needs to happen in an emergency. Managing human resources relies on specialists in HR offices as well as the leadership found in the owners or executives for any level of business. People drive business success and caring for the people in one's workplace should always be prioritized as part of a company's business ethics and business continuity plan.

Employers should work closely with their human resource management team (HRM) to reduce physical, economic, and psychological impacts and to design how to step in and provide resources for those affected. Employers also need to care for their leadership and for themselves, because mass emergencies carry a heavy burden to those tasked with making difficult decisions that affect both people and the business. In short, human resources and HRM remain key to business survival in a disaster (Mann & Islam, 2015). This chapter first reviews what happened to human resources in several kinds of disasters to sensitize readers to the value of human resource centered planning.

Case examples

To ensure broad coverage of the unique issues that can emerge, three different types of events will be used to illustrate three kinds of disaster: human-made, natural, and viral.

Terrorism, September 11 – The United States of America

September 11, 2001 brought significant impacts to human resource professionals and employers both in the U.S. and, because of the attack sites and

Business Continuity Planning. https://doi.org/10.1016/B978-0-12-813844-1.00008-7

employees, around the world. Certainly, the loss of life represented the greatest impact and challenge to those tasked with human resource management. While HR professionals are accustomed to assisting employees in a crisis that generates bereavement, this one was different. Working through the tragedy meant dealing with personal and professional losses while also navigating the unspeakable grief that consumed people worldwide. The disaster also occurred in multiple locations and within businesses that operated globally – the combination would prove daunting to even the most experienced HRM professional.

In New York City, businesses occupied entire floors and lost massive numbers of employees when the planes hit the World Trade Center. Cantor Fitzgerald, a financial services and securities firm, lost 658 employees who worked on the top five floors of the World Trade Center. The result: a catastrophic loss of two-thirds of the New York area office that affected 30 affiliated offices located around the world. Other businesses lost hundreds of employees too, such as those working for Marsh where 295 employees died. New York City lost 343 of its firefighters and 71 police officers when the towers collapsed. In Washington, DC, both civilian and military lives were lost in the attack on the Pentagon with critical losses in command officers who normally supervise people in an emergency. In the Pentagon, 125 died including 70 civilians. The plane that hijackers crashed into the Pentagon included 64 souls on board. The plane that went down in Pennsylvania, when passengers bravely attempted to retake control from the hijackers, carried people involved in many types of businesses. Forty passengers, from four different nations, perished.

In New York, communication emerged as the first challenge once the buildings collapsed and survivors emerged – who had survived? For weeks, employers waited in agony with their personnel among the missing. Employers could not find – nor easily get in contact – with their surviving employees. In the days that followed, human resource professionals worked to help those affected with insurance coverage for injuries, job losses, mental health impacts, workplace displacement, emergency leaves, and unexpected travel for response and recovery efforts but without the building they operated in to do that critical work. Thousands of workers outside of the affected buildings, particularly in the New York travel and hospitality industries, lost their incomes. Simultaneously, workers had to be found – and protected – to clean contaminated offices in the immediate impact areas. Insufficient protective equipment meant that those workers, particularly at the lowest pay levels, faced significant consequences for their health and safety (Lippmann, Cohen, & Chen, 2015).

But Americans moved through this tragedy and began to restore business operations and support affected employees. Insurance helped, with approximately $40 billion in claims paid out for business interruption, building damage, worker's compensation, health, and injury issues. Some businesses reduced employee numbers while others leaned into new opportunities, including the emerging field of homeland security (Castillo, 2004). Nearly 20 years later, however, lingering health issues remain for those who survived, from a debilitating cough to cancers related to contaminants (Lippmann et al., 2015). For those affected, government officials created multiple victim compensation and health registry funds to address long-term health effects, including those that caused terminal cancers.

Natural disasters

Natural disasters sometimes occur without warning, like an earthquake. Others take time to unfold, such as the days before a hurricane makes landfall. Both present significant human resource challenges. In early September, 2010, a 7.1 Richter magnitude earthquake struck the Canterbury Region of New Zealand, on the nation's southern island. The earthquake weakened buildings, but most remained usable. On 22 February 2011, however, a 6.2 earthquake severely damaged the Christchurch central business district. Buildings collapsed in a concentrated area, causing 185 deaths as well as damage to 3000 buildings. Within the affected central business district, 70% of the damaged buildings displaced 50,000 employees (Hall, Malinen, Vosslamber, & Wordsworth, 2016; Malinen, Hatton, Naswall, & Kuntz, 2019). The total cost for physical damage would reach nearly 3 billion in New Zealand dollars. The heaviest concentration of fatalities occurred in a single building called the Canterbury Television Building or CTV. In addition to the area's television setting, the building included a medical practice and an English as a Second Language (ESL) school. Half of the deaths occurred in the CTV with 16 media employees, 19 medical patients and staff, and 71 foreign students. Twenty nations lost citizens in this building alone.

In a more slow-moving disaster, hurricanes Katrina and Rita severely damaged the U.S. Gulf Coast in 2005. Impacting a massive area, the size of the United Kingdom, the damage left millions of people displaced from their homes and workplaces. Hard-hit areas in lower Louisiana included the City of New Orleans which became a near ghost-town for much of the next year. Lacking utilities in most areas, businesses sat vacant while employees transitioned from one temporary housing situation to another. Even essential businesses had to work remotely, many from the city of Baton Rouge about a 1-h drive from New Orleans. The central tourist area, called the French Quarter, suffered wind damage and minor flooding and was among the first places to re-open. Other businesses re-opened slowly as resources and residents returned, but only after the U.S. Army Corps of Engineers "de-watered" the city from a levee system broken by wind-driven storm surge. The city re-opened slowly, with a restaurant here and a coffee shop there, amid the dull brown of deadened landscaping. Small, handmade signs announced slowly unfolding re-openings that continued for years. Many businesses did not survive. Federal and state funding and loans fueled some rebuilding, with micro-loans enabling smaller enterprises to return.

On the Mississippi coast, small and medium-sized towns suffered heavy devastation from a massive storm surge that pushed ashore and damaged the fishing industry, tourism including beach activities, gambling, hospitality, and restaurants. Small towns that had been home to artists, jewelers, and writers faced a daunting return through roads and highways clogged with debris. When business owners did make it back to the coast, they returned to a landscape scoured by the storm surge. Police departments, libraries, and governments had to rethink how they would provide services without facilities. In one town, officials held formal meetings under one of the few trees left standing – because city hall was gone. Employers, elected officials, and essential workers set up operations slowly in mobile homes or trailers as they became available. Libraries established mobile operations and used alternate locations while their facilities were cleaned, decontaminated, repaired, or rebuilt. The recovery took years, fueled by the adaptability and dedication of employees and government (French, Goodman, & Stanley, 2008).

Pandemics

In 1918, a global influenza outbreak infected approximately 500 million people and claimed around 50 million lives. The outbreak spread worldwide and occurred in three waves from the spring of 1918 into the summer of 1919. Health care systems were overrun, workplaces shut down, and people engaged in new forms of behavior from social distancing to walking to work to avoid public transportation. Occurring at the same time as World War I, military members spread the illness due to necessary troop movements, debilitating those essential to global security. Workers fell ill everywhere, resulting in closed businesses that now lacked enough employees. To restore services essential to people (food, fuel, military support, health care) workplaces staggered shifts (CDC, 2020).

Everyone experienced the 2020 coronavirus pandemic – every business, every employee, every nation – and in many ways. This one was both personal and professional everywhere, and strategies to deal with the global outbreak reflected 1918 strategies over 100 years later. Fear, anxiety, and stress characterized the impacts on employees and employers particularly among health care workers, first responders, grocery workers, and delivery services. With "stay at home" and "work at home" strategies required to reduce the spread of the virus, businesses failed at record rates and nation after nation moved to address massive unemployment. As the pandemic went on, furloughs, salary reductions, and layoffs surfaced as businesses continued to fail or declared bankruptcy. Human resources offices proved vital as they supported staff transitioning into telecommuting, experiencing anxiety, or going through a job loss.

The critical role of human resources management

In each of the scenarios discussed here, employees and employers required the support, knowledge, and guidance of human resource management professionals. Most businesses centralize human resources management to promote consistency and efficiency, thus they represent a crucial asset during an emergency (Premeaux & Breaux, 2007). The personnel in these offices manage recruiting and onboarding new employees, address employee issues such as payroll or benefits, handle leave through explaining and overseeing related policies, and support family members when their working member has been affected. As realized after the hurricanes Katrina and Rita in the U.S. in 2005, human resources management professionals prove to be a critical and stabilizing force in a disaster (Goodman & Mann, 2008).

Employers and employees rely on human resources personnel at a high level, thus involving them in a business continuity plan should begin at the earliest stages of planning. HR professionals will see and know issues that other planners will not realize – like the range of policies that people can use when disruptions occur for matters like telecommuting, flex time, and hazard pay. HRM offices also hold vital records essential to contacting the workforce and can help to ensure that shifts are covered by employees qualified to do specific work. They also serve in an emotional capacity because they know how to offer information to employees under stress or dealing with unforeseen circumstances. Human resources personnel thus represent a critical workforce element to retain and rely on – and to resource adequately so they can do their job.

Business continuity planning for human resource management

It is imperative that HRM professionals become deeply involved in crafting business continuity plans. They will need to participate in writing one for their own specific unit so that it remains functioning and in the larger enterprise's overall plan. HRM professionals must first draft their own critical functions specific to how their unit works. HR planners will then need to consider where they will do that work, which in a disaster might require relocating to a new command center, working from home, or staying onsite throughout a difficult time period (Goodman & Mann, 2008; Henry, Cho, & Dupuis, 2008; Premeaux &

Breaux, 2007). Depending on the size of the HR unit, planners will need to consider how they will then address the critical function. Who is an essential worker in each critical function and what is their role? What are their work-arounds considering potential direct and indirect effects? For example, how will HR deal with managing shifts or processing payroll should power be out? How will HR personnel access, create, and safeguard confidential employee records from a remote site? How will they hire and onboard new employees in an emergency or handle a hiring freeze? What resources do they need to address the critical functions like hiring, training, benefits, and payroll? How will HR manage employee emergencies (e.g., injuries, medical or family leaves) in a disaster? Which agencies can supply temporary workers, and can they do so in an emergency? These questions must be answered in advance to make things run more smoothly and need to be specific to the business's values, priorities, and policies.

Second, HR needs to work with management to address how business policies will address disaster-time conditions and how work functions might shift to new needs. For example, *downtime* (see Chapter 2) occurs when a business cannot function – meaning that employees will not be working. HRM will need to work with management in advance to create policies how they will pay – or not pay – workers under downtime conditions. If the assembly line is shut down, will workers be paid? Does a union contract cover such conditions? Will the disrupted revenue stream permit payroll distribution or will business interruption insurance help? Disasters can also reduce the available workforce, which happened with hurricane Katrina. Pascagoula, Mississippi lost half of their employees amid a significant demand for increased services (Goodman & Mann, 2008). Given the massive destruction that destroyed or damaged thousands of homes and workplaces, employees could not return to work and HR had to launch job fairs. Those who could work during the Katrina recovery soon faced problems securing their paychecks. Without power, some workplaces relied on generators to do payroll – but with banks affected, they could not do direct deposit nor could employees access their pay (Goodman & Mann, 2008). Business continuity planning tasks HRM professionals with anticipating and solving this type of problem. For example, some areas subject to repetitive hurricane strikes issue a dummy payroll before severe weather arrives, which can be cancelled if the storm turns direction (Goodman & Mann, 2008). Not surprisingly, communications about pay checks emerge as one of the most important messages that HR can address after a disaster (Nilakant, Walker, Rochford, & Van Heugten, 2013).

Similarly, *displacement* (see Chapter 2) may occur in various ways from telecommuting to working in a completely new business location. HRM offices can help with the transition of employees to a new location and determine who can – or cannot – work from that site. Disasters disrupt public transportation and damage roadways used by those driving personal vehicles and employees may not be able to get to a new worksite, which occurred in the 2017 California wildfires (Massarweh, 2019). Thus, displacement may involve HR in finding temporary employees or advocating for alternate transportation resources for existing employees (Woodward, 2006). After Katrina, HR services also included finding temporary housing for workers who lost their homes, by inventing a Share a Home program (Ladika, 2006). Clearly, the consequences of a disaster can be significant for both employees and the human resource team that supports them. Having a business continuity plan enables HR to foresee the potential consequences of a disaster and be ready to pivot quickly and effectively when needed.

Preparing employees for disruptions

HR professionals can be even more pro-active by encouraging employees to be ready for a disaster. For example, a significant business challenge occurs when an employee cannot work because of a lack of transportation, illness or injury, or damage to their home. Who covers that employee's critical functions? One way to manage temporary personnel losses comes from cross-training, where employees learn each other's jobs. People within similar units could train on procedures necessary to do each other's jobs or just spend time observing another person. That way, if someone is out sick or injured, businesses can manage critical functions and be better positioned to survive.

HR could also sponsor an annual disaster preparation day where everyone focuses on their critical functions. Just practicing could identify and prepare employees for what they may face (Goodman & Mann, 2008). They may face significant worry and anxiety about fulfilling their critical function depending on the emergency. For example, essential employees who handle the mail or clean a building have faced anthrax sent through the mail (five employees died), encountered unknown substances, and dealt with coronavirus-contaminated packages and surfaces (e.g., see Blanchard et al., 2005). Disasters also take away both heat and air-conditioning, causing employees to work in power-loss situations where heat and humidity or bitter cold can debilitate workers quickly both inside and outside of workspaces. Walking through emergency procedures, which should be required training, can help to reduce that anxiety and surface procedures and policies to protect the health, safety, and welfare of employees.

Similarly, in anticipation of a work disruption, HR can advocate that workers expected to telecommute should routinely practice doing so. Practicing will enable the employee and their unit to identify trouble spots, such as connectivity, remotely accessing a shared drive, dealing with critical data not allowed to be downloaded or viewed offsite, or similar problems. And even telecommuters face challenges at home when the Internet drops out or a tree downs a power line. Telecommuters should thus be ready at home to shift their critical functions to a colleague via their own backup plans. By practicing, telecommuters will be able to step into their roles faster and the business will be able to transition or adapt more ably.

By preparing employees, HR helps to generate a "new normal" more smoothly. For hurricane Katrina, the new normal meant that universities operated out of mobile trailers, hotel rooms, and coffee shops. Businesses in New Orleans set up remote command centers in Baton Rouge and shuttled employees in and out. They also provided temporary housing, organized flex time, and created leave donation programs (Goodman & Mann, 2008; Henry et al., 2008; Premeaux & Breaux, 2007). Wouldn't it be easier to design this in advance through an HR-centered business continuity plan than dealing with it during a disaster?

What employees may experience

Depending on the disaster, employees may face a range of impacts including downtime, displacement, and disruption. Those effects tend to fall in to two different situations, with personal and professional consequences. HR needs to anticipate these impacts and to be ready to support their workforce.

Personal impacts

The first consideration of any employer should always be the well-being of their employees. Any disaster has the potential to cause harm to an employee. An active attacker can cause injuries and fatalities as can a building that collapses in an earthquake. In such circumstances, caring for those injured becomes of paramount importance from providing insured coverage for medical procedures to offering medical leave. Another element also comes in as the disaster unfolds, with the effects on families of those injured. HR may need to arrange for family/caregiver leave so that an employee may care for a family member who was injured or who became ill. In a disaster, employees will also experience non-disaster emergencies that require support.

Relevant help may emanate from an Employee Assistance Program (EAP) or by creating a list of recommended providers. Many EAPs today offer caregiver respite options, referrals for home health agencies, and mental health support. Workplaces may also arrange for care teams (formally or informally) to support employees while they or their loved ones are going through treatments and recovery or as they repair damaged homes. Care teams might also provide transportation, temporary housing, and meals or connect the employee with such resources. Longer-term care teams might focus on transitioning a home environment to be accessible and supportive when an injured employee comes home. In the worst disasters, care teams might connect with voluntary organizations to repair or rebuild the homes of people who experienced direct hits. Often such care teams emerge out of employee volunteer efforts, but a good business continuity plan encourages care teams to be pre-developed. Even if they are not in existence before disaster, it is likely that people will want to help their co-workers. For hurricane Katrina, HR professionals organized an Adopt a Family program to provide support (Ladika, 2006). Disasters also cause fatalities, understandably the most difficult employee situation to address. HR will need to work with the families to arrange for final pay checks, bereavement support, life insurance payments, and the return of personal possessions left at the workplace. Such care responsibilities may take a toll on HR employees who will want to build in a process to care for their internal workers.

As recovery ensues, many workplaces and communities create events to honor those who stepped up and to recognize any employees that died. Many such events occur at the 1-year anniversary of the earthquake, tornado, or flood, a marker in time that enables those affected to look back and realize how far they have come (Eyre, 1999, 2006; Richardson, 2010). Events that commemorate losses include formal services that memorialize the fallen, such as the annual remembrance of firefighters held annually at fire stations on September 11. Less formal gatherings may involve candlelight or silent vigils, fairs that celebrate local cultures, or personal gatherings (Eyre, 2006; Forrest, 1993). Communities also memorialize events with clothing like t-shirts, posters, pins, jacket patches, certificates, awards, extra days off, and bonuses. Pausing to remember and recognize those affected, those who were lost, and those who survived matters to people and should be part of human resource planning.

Professional impacts

Disasters clearly cause significant impacts to people's livelihoods and professions. With buildings lost or homes destroyed, people may lose their workplaces owned by others or their own home-based work locations. Many small business owners – carpenters, plumbers, insurance agents, financial analysts – and so many more – may work from home. In a tornado strike, one's tools, inventory, computers, and records can be lost in seconds.

The disaster may also imperil an entire industry's base, as the 2004 tsunami did with coastal fishing industries or when the 2010 Deepwater Horizon oil spill affected the Gulf of Mexico. The tsunami damaged marine and land environments, reducing fishing harvests and increasing salinity in agricultural areas (De Silva & Yamao, 2007). Destruction to marinas, markets, and fishing resources further decimated people's abilities to overcome profession-related impacts. For the Deepwater Horizon event, the spill affected fisheries, family businesses, and the supply chain from fishing boats to markets. To make things harder, the disaster came after 5 years of dealing with hurricane Katrina followed by the economic recession of 2008 (Lambert, Duhon, & Peyrefitte, 2012; Phillips, 2013). People in Louisiana kept having to react to one disruption after another.

People may have to rethink where they will secure their livelihoods after a disaster. That was the case in the 2020 pandemic, when record numbers of employees faced short-time unemployment followed by business failures that resulted in longer-term unemployment. In the U.S., the government moved to provide economic stimulus funding and moved people onto the unemployment rolls where they could secure additional support. The U.S. based Small Business Administration also offered a paycheck protection program based on loans to affected businesses with under 500 employees. In the United Kingdom, the government paid up to 80% of lost wages for up to 3 months. The Netherlands pledged up to 90% coverage for lost wages. Denmark paid between 75% to 90% of worker paychecks. Such a significant livelihood impact is rare in a disaster, as are the programs that were introduced. More commonly, employees may have to retool for a new job or career, or human resources may need to train employees as the business adapts. Companies that did so during the pandemic adapted manufacturing capacities, such as transitioning from building cars to hospital ventilators. For the Deepwater Horizon oil spill, those employed in the fishing sector found temporary employment cleaning up beaches, rescuing wildlife, and laying barricades to prevent oil intrusion into wetlands. Disasters also generate other opportunities to rebuild and repair homes and businesses, organize temporary housing units, or supply critical materials. It may be that HR will need to redirect employees to opportunities that provide income while the business regroups.

What employers may experience

Employers also experience significant challenges when disasters strike, beginning with how to address critical functions given the potential employee impacts. This section thus starts first with the people who make up the business and how employers can address losses and impacts to their workforce.

Coping with employee impacts

Because revenue disruptions undermine a business's ability to pay employees, an owner may need to decide quickly who they can pay and to sort through various options. Those options can be pre-determined by the loss estimation (see Chapter 3) that revealed when disaster effects and costs outpace revenue streams. During planning team discussions over loss estimations, decisions can be made about the breaking point where businesses balance revenue versus expenses. That discussion will influence options for employers and their employees which may need to include:

- *Reduced working hours.* In this option, employers cut back on the hours that employees work to conserve revenue but still enable employees to earn some income and the business to meet critical functions. Alternatives to lost payroll can be launched simultaneously. In the recent pandemic, employees in the bar and restaurant industry launched fundraisers for colleagues who were negatively impacted or collected tips for use by affected employees. Reduced working hours can also be offered across the workforce, as some employees with children or caregiving responsibilities or those nearing retirement may prefer an hourly reduction to manage their personal and professional lives.
- *Move to essential employees only.* Some businesses may need to reduce staffing to only essential employees needed to operate the physical plant or conduct specific critical functions. Employers may or may not opt to pay those not deemed essential workers. For those designated as essential workers, employers should consider hazard pay and always provide personal protective equipment and disaster-relevant health and safety measures. In some situations, governments may designate which businesses are deemed essential which may also influence whether a business can remain open.
- *Offer transfers to new or alternate locations.* Larger chains and corporations may be able to move employees out of devastated sites and to other locations from which they can work or offer telecommuting where feasible. Similarly, employees from outside the affected area maybe able to bolster locally impacted workforces when the number of available employees becomes reduced.
- *Create a donated leave pool.* Donated leave could be used to keep people on a payroll. In the U.S., IRS Notice 2005-68 allows for employees to offer their available leave for a pool that employees can use in a disaster. Such an option may be helpful for those dealing with their own personal household recovery or in situations where employees must be taken off a payroll.
- *Furloughs.* Employers may need to furlough employees which means that they have a job they can return to but will not be paid during the furlough. Typically, employees retain their benefits such as health care. Furloughs exist for a specified time and may unfurl slowly across the entire employee pool to spread out individual impacts.
- *Hiring freeze.* Many businesses put hiring and purchase freezes into place when a disaster happens to conserve available funds. Such an action should be taken as soon as possible after a disaster to maintain business viability.

- *Reductions in force (RIF)*. Employers may choose to downsize the number of employed workers when the business experiences significant financial losses. RIF processes mean that those who fall into the reduced category are unlikely to return to their job. Some employers may offer incentives to make the process less painful such as Early Retirement Incentive Programs (ERIP) that provide a financial payout or extended benefits to those who want to leave the workforce voluntarily.
- *Reorganizing*. A business may opt to reorganize their structure to streamline operations by combining positions or centralizing functions. That reorganization could result in layoffs or reductions in force. In some instances, businesses may choose to re-hire employees into the newly created positions.
- *Layoffs*. When industries face the reality that revenue now outstrips expenses, layoffs may be necessary. Layoffs may or may not hold the possibility of a return after a specified time. Employers will likely find that when they try to bring employees back, they will face a loss of talented employees who have sought other positions to support their families.
- *Declare financial exigency or bankruptcy*. In many locations, businesses can choose to declare bankruptcy as a way of protecting their assets, although such an option is not without subsequent challenges. Another step is to declare financial exigency which may afford opportunities to reduce staff that would normally not be released. Financial exigency is used in higher education which allows a university to release tenured faculty. Such a measure is considered drastic as tenure represents an earned, lifetime guarantee of employment.
- *Support programs*. In many locations, particularly developing nations, it may be possible to pursue micro-loan programs or offer cash for work/food for work. (Srivastava & Shaw, 2015). The goal of a micro-loan program is to give out a small amount of money so that a business can secure resources to produce useful goods. After one of Turkey's earthquakes, local women producing dolls at a relief camp secured a small grant from the ministry's tourism division. They converted the grant into a successful doll production factory providing pay checks for displaced workers (Yonder, Akcar, & Gopalan, 2005). A similar micro-loan program after hurricane Katrina in the U.S. supported small, minority owned businesses often located in people's homes (Phillips, 2015).

Coping with immediate business impacts

What should a business owner do first? In past disasters, employers have always taken care of their people first. That means attending to conditions that affect life safety, which relies on effective communication of risk and the appropriate actions to take in specific kinds of disasters. HR and IT should be tasked with insuring that they know who to contact, how to contact them, and to have back-up strategies in case a first or preferred mode of communication fails. Employees also need to know where and how to get information, so a streamlined and focused education plan on how to check-in and acquire more information is essential. Many businesses use email, social media, and new web pages to communicate. The reliance on technology should raise eyebrows, though, and planners will need to designate alternative ways to communicate – because power outages and cyberattacks can render such tools less useful.

Next steps will likely involve supervisors and owners in assessing the physical space of a building, its infrastructure including technology hardware, software, and connectivity, and damage to contents. If the building is safe to enter, what immediate steps need to be taken to salvage the business such as a tarp roof to reduce water intrusion or the removal of key resources like computers or cash registers? Building assessment should be undertaken by professionals, and governments often do such damage assessments. Insurance companies need to be contacted to cover losses as well. Professional disaster recovery teams can be brought in, though in a major disaster it may be difficult to secure them. Thus, prior arrangements should be made for specific needs. To illustrate, removing water and drying out a flooded building can reduce longer-term issues including water absorption into sheetrock and insulation or mold growth. In the business continuity plan, pre-identifying vendors who can assist with the more higher probability disasters would be advisable.

Businesses then typically move directly into restarting with an emphasis on critical functions, which should have been prioritized in the business planning process along with an appropriate timeline. In the case of human resources, employers need to work with their HR team or (for small businesses in particular) directly with employees. The BCP will have outlined the critical functions that teams of employees need to undertake to restore the business into operational status. Employers will need to lead or coordinate those teams with regular meetings that surface problems, identify resources, and resolve issues. Restoring critical functions serves as the first step toward recovery, which was the goal of all the previous efforts taken to create a business continuity plan.

Navigating recovery

The response phase of disasters concentrates on saving lives, providing for life-saving resources like food and shelter, preserving properties, and picking up the pieces. Recovery, which lies at the heart of business continuity planning, takes much longer as families, businesses, and communities restore infrastructure and utilities, begin repairs and reconstruction, and deal with the aftermath of losing homes, jobs, and loved ones. Depending on the impact or duration of an event, recovery could play out from days to years. Employers will need to be ready as they move employees and their business along the road from life-saving activities into business restoration.

Transitioning from response to recovery

In the Christchurch earthquake, initial needs included food, shelter, and water. Businesses simultaneously focused on accounting for their employees followed by assessing family needs, which informed employers about employee availability and related stress levels (Nilakant et al., 2013). Affected businesses learned to do careful assessments of employee's home situations because employees often under-reported earthquake impacts, assuming others had fared even worse (Nilakant et al., 2013). Determining impacts can be difficult in such circumstances. In New Orleans after Katrina, it became commonplace to ask how much water someone had in their homes. The answer – a few inches, a few feet, or 8 ft (the ceiling and beyond) influenced when an employee could return to their homes and thus to work and indirectly suggested potential stress levels. Yet even as affected employees finish their home repairs, so does their workplace. Many employees return to work physically and mentally depleted from what they had to deal with at home only to become more exhausted restoring the business.

Yet people often feel they cannot share what happened or just do not want to discuss it any further. Pennebaker (1990) reported that when people stop telling their earthquake stories, higher rates of headaches, backaches, and psychosomatic illnesses surface. In Christchurch, earthquake-affected employees reported higher rates of head colds and increased performance errors (Malinen et al., 2019). Workers also reported stress due to new working environments and equipment coupled with national resource reductions and general job insecurity (Malinen et al., 2019). Managers, who felt they took on new roles to counsel and support supervisees, reported higher stress levels. Employees reported being completely exhausted before a year had gone by (Nilakant et al., 2013). Employers must therefore attend to physical and mental health impacts to improve work productivity and to retain employees (see Box 6.1).

In some post-disaster environments, employees may be hard to find because of evacuations or homes lost to the event. To populate recovery, employers may need to recruit and hire employees and may face competition to find qualified workers. After Katrina and Rita, the states of Louisiana and Mississippi respectively lost 256,000 and 60,000 jobs coupled with a massive dispersion of evacuees across the nation. Re-hiring employees proved difficult and companies, agencies, and government offices had to offer a higher hourly rate, housing,

BOX 6.1 The psychological impacts of disaster.

Given the potential personal impacts, employers would be wise to anticipate mental health impacts ranging from the more commonly experienced stress to the seriousness of post-traumatic stress disorder. However, in a meta-review of studies involving over 60,000 individuals, researchers found that most of the time people respond with a fair degree of resilience (Mankin & Perry, 2004; Norris, Friedman, & Watson, Byrne, et al., 2002; Norris, Friedman, & Watson, 2002). The good news, then, is that people can and will rise to the occasion with resilience. Employers can help to promote that resilience through addressing psychological impacts and offering appropriate programs and resources.

As noted in the Christchurch example, stress may be a commonly shared employee experience which affects worker performance. Direct exposure to the disaster, particularly closeness to injuries and fatalities, may worsen mental health impacts if not addressed effectively (Paton, 1999). Even professionals who deal with emergencies daily require additional training to prepare for a massive event and benefit from post-disaster counseling services (e.g., see Brondolo, Wellington, Brady, Libby, & Brondolo, 2008). Survivor grief and shock may occur (Castillo, 2004).

Workplaces can involve mental health professionals in offering stress relief activities from physical movement (e.g., yoga, break times to walk, stretching at your desk) to more social interactions. In Sweden, social interactions occur twice daily during normal times and are called "fika." At the fika hours in the morning and afternoon, co-workers gather for coffee and a snack which facilitates short interactions (Uusimäki, 2020). Such regularized social interactions promote connections that reduce isolation and solidify social connections. Social networks and social ties, generated through that interaction, can prove helpful in building people's resilience to disaster impacts. Even when people work virtually in a pandemic, connections within the home and across cyberspace can help. Upon returning to work, employers are well-advised to offer social times when people can reconnect in person over breaks, potlucks, or other kinds of connections (Nilakant et al., 2013).

Helping employees with mental health and general health care needs may be difficult in a post-disaster context. After Katrina and Rita, employers reported that available medical care remained meager and "primitive" (Henry et al., 2008, PAGE). Limited mental health resources existed. In the COVID-19 pandemic, telehealth emerged as a viable option for the delivery of both physical and mental health services. The strategy served to mitigate spread of the virus and to provide health access across significant distance, for those with insurance and a means to communicate via telehealth options.

Because "employers must endure basic psychological security of employees as a necessary foundation for organizational performance" providing mental health support is crucial to do (Mankin & Perry, 2004, p. 15). Preferably, employers will provide appropriate training prior to a disaster (Brooks, Dunn, Amlôt, Greenberg, & Rubin, 2018). Employee Assistance Programs (EAPs) often deliver those mental health and other resources. EAP services increased after September 11, as employees navigated new kinds of anxiety related to worker safety (Mankin & Perry, 2004). The same concerns resurfaced for COVID-19 as employees navigated returning to a physical workspace amid a still-present threat to public health. Workplaces with EAPs made services available via telehealth or video conferencing systems and many offered stress reduction activities including breaks, walks, and yoga.

Supporting employees through a traumatic and exhausting experience includes care for the executive team as well. Long days and nights, limited time on weekends or traditional holidays, and a continual pressure to balance critical functions with an eroding financial bottom line can enervate even the most experienced executives. Further, executives have a responsibility to model appropriate behavior for employees. If the boss does not take care of herself and those around her, then her employees may feel similarly.

transportation, and childcare to compete (Woodward, 2006). To help with the recovery, the U.S. government relaxed the grace period for international work documentation (Woodward, 2006). But such a strategy will not always work in every disaster. Closed borders during the COVID-19 pandemic led to several nations losing major crops because of a lack of workers.

Recovery can take hours or years. Many variables influence re-opening including access to the site, which is compromised by many kinds of disasters. Re-opening also relies on the availability of infrastructure to support operations, from transportation arteries to supply chain routing to utilities. Employees will be needed yet may not be available as indicated earlier in this chapter. Downtime, displacement, direct, and indirect impacts will also influence the time to return as will the general state of the economy prior to the disaster (Wasileski, Rodríguez, & Diaz, 2011; Webb, Tierney, & Dahlhamer, 2002). Business as usual may not return for some time, with businesses needing to adapt, downsize, reconfigure, regroup, and redetermine what owners must do first. Business continuity plans inform which critical functions must be undertaken to relaunch operations, coupled with the procedures, teams, vendors, and resources they need to make it happen. In this section, readers will look at several reopening examples with an eye on the challenges faced by each, the context in which it happened, and the resources required to begin again.

Natural disaster: The 2004 tsunami

The 2004 tsunami affected 13 nations, heavily damaging fishery operations, small business sectors, tourism and hospitality businesses, home-owned enterprises, schools, government, and the nonprofit sector. Within India, a small ten kilometer stretch of shoreline sustained significant damage, with as many as 18,000 souls perishing in the waves. The community of Vailankanni, on the southeastern sector of the nation, included a large commercial sector stretching out from the beach and slightly uphill to Our Lady of Good Health Basilica. The tsunami wave, described by survivors as over 40 ft high, scoured the commercial sector stopping just short of entering the Marian shrine.

Vailankanni's commercial sector benefited from the area's agricultural and fishery operations as well as the tourists who visited the Basilica. External organizations took that broader context into consideration when they began re-opening the area, by thinking holistically and in concert with local, traditional practices (Oxfam, n.d.). Outside organizations first focused on shelter, food, and emergency care for survivors, the majority of whom had also lost their homes and family members. The local Basilica donated land for mass graves and launched programs to feed people and provide spiritual care. Temporary shelters went up, with short-term work programs inside them. Cash programs hired cleanup workers to manage the debris and created women's groups to convert recycled materials into sellable products. With businesses that relied on agriculture and fishing, non-governmental organizations provided boats, refrigeration trucks, docks, and fish stalls to restore the livelihoods that depended on each step in the sea-to-market supply chain. The commercial sector thrived again.

The Loma Prieta earthquake

Earthquakes cause damage that unfolds for some time, as engineers and experts assess damage within the walls of buildings or in underground utility systems. The 1989 Loma Prieta earthquake, which struck an area from Santa Cruz and Watsonville to San Francisco, did just that. The Pacific Garden Mall, located in downtown Santa Cruz included blocks of small, unique businesses reflective of the local community. Patrons loved going downtown, where they bought Halloween costumes from the same retail store every year, purchased freshly roasted coffee, and browsed a favorite bookstore. Several buildings served as single room occupancy homes for low-income residents. Tourists visiting the popular beaches and surfing areas shopped there as well, and enjoyed a small theater, unique restaurants, and historic architecture characteristic of the 1920s. But the outdoor mall rested near a fault line that shifted violently on October 17, causing major damage to downtown businesses. Officials assessed the damage and roped off the entire central area. It would be years before they navigated the damage, which meant tearing down badly damaged buildings, seismically reinforcing others, and repairing broken underground utilities. Streets had to be made safe as well, and debris that fell from the two and three-story buildings had to be removed or carefully stored for re-use.

Within days, however, business owners, community members, and the city moved to restore customer traffic and keep businesses viable. Signs went up indicating which businesses were open and added directional arrows to guide customers to their doors. The city partnered with businesses to establish tented areas with small booths, called the Phoenix Pavilions, to inspire the idea of rising from the ashes. Community residents turned out to hand carry boxes of books to a temporary site for the favorite bookstore, newly designated the Book Tent Santa Cruz (see Box 6.2 for more about ties between businesses and communities). Within months, businesses were generating income albeit less than before. The city next convened an effort called Vision Santa Cruz which invited stakeholders to design goals for a return to normal. Within the first year after the earthquake, that effort established a downtown office where anyone could walk in, view examples of damage as well as options for the future, and offer their opinions. A series of planning charettes revealed community favorites, which emphasized a return to the historic character and feel of the downtown sector. The Small Business Association made loans to business owners while other grants and loans fueled repairs. In short, the Santa Cruz downtown recovery required a wide set of partnerships contributing to the vision and funding of business recovery. Within a few years, many businesses had returned to the downtown area which became the site of annual memorials. The downtown Pacific Garden Mall thrived again, because of how the community became so involved. It's the people who make business recovery possible.

BOX 6.2 The business and the community.

Disasters bring out the best in people and employers are part of that spirit of rising to the occasion. Businesses always step up in a disaster, from feeding massive numbers of people (Andrés, 2018) to organizing smaller-scale fundraisers and donation drives. Communities rely on businesses, and employees often feel better about the situation when they can contribute to helping others. Such altruistic responses have come to characterize what we see in disasters, and even when businesses have been negatively affected, employers and employees step up time and again. The COVID-19 pandemic dramatically undermined the financial viability of numerous enterprises – yet those same businesses used their resources to feed the community, make or donate personal protective equipment, and offer consulting services at no charge.

The relationship between a business and the community matters because customers, employees, and clients are watching. When a business pivots to serve its community, it generates goodwill and positive images that can carry the business into and past the disaster. Community members will also want to support businesses that lead efforts and will lean into new opportunities to make a difference. Part of that contribution can be adapting resources to new ways of helping (see Photo 6.1). A company that relies on interstate transportation might use its vehicle fleet to deliver donations. Small financial services might shift into helping newly unemployed workers redetermine their household budgets and how to conserve available funds. Services that provide personal transportation might shift to taking injured survivors to health care settings for medical appointments or physical therapy. Schools can adapt to closures by continuing the delivery of food to hungry children. Universities can extend expertise to track the effects of disasters and make projections that inform governmental response. The faith sector can organize rituals that recognize grief and support survivors. There is a role for every kind of enterprise to contribute to its community.

PHOTO 6.1
Businesses pivoted productions to make ventilators for response to COVID-19. The effort leveraged business capabilities in the states of Washington and Indiana. *Jeff Markham/FEMA News Photo.*

Recognizing and rewarding employees

Following the Christchurch earthquake, workers needed affirmation as they diligently took on increased workloads in new settings and with different resources (Malinen et al., 2019). Businesses can and should organize ways to recognize their employees. Strategies might include:

- Using web pages and social media to recognize those who have made a considerable difference.
- Giving awards to employees who have gone above and beyond or made a significant contribution to the disaster response and recovery effort.
- Offering certificates, monetary awards, or plaques to employees to recognize the hard work they put in during the disaster.
- Displaying photos on office walls, websites, or social media to alert the broader community to the value that their employees provided.
- Creating a special pin, patch, or shirt around the specific disaster that can be given to each employee who went through the event and supported their workplace and/or the broader community.
- Taking the time to thank every employee individually and in person. Where this is not possible, create a video to share with everyone.
- Organizing a recognition banquet or collective pot-luck where people can gather to thank each other and to reunite.
- Putting up billboards around the community or placing an ad in the paper recognizing and thanking their employees and the community, clients, and customers who helped them to survive.
- Erecting signs at the business in its landscape or windows to thank everyone – employee, customer, or client, for all they did.
- Nominating employees and businesses for awards and recognition outside of the company, especially for what they did to give back to the community.

Essential actions

For this chapter, business continuity planners should take several essential actions:

- Meet with or include the human resources department/personnel to discuss their involvement in the overall planning effort.
- Identify critical functions specific to human resources as configured in your organization.
- Identify HR personnel who will bear responsibility for carrying out the critical functions specific to employee support and business survival and create work teams for those critical functions.

- Organize a workplace care team that will address small to large emergencies, including support with formal benefits and informal, voluntary help.
- Specify an alternate workspace from which HR can set up operations quickly and create a "go kit" they can take with them to do that work.
- Organize a website section specific to disasters and emergencies for employees and link it to available resources and pertinent policies.
- Pre-plan key support activities that may be needed after a disaster such as transportation, housing, childcare, hiring and onboarding, and training.
- Ask HR to review all planning efforts to date from their perspective of providing employee support in a crisis.

References

Andrés, J. (2018). *We fed an island.* New York: HarperCollins Publishers.

Blanchard, J. C., Haywood, Y., Stein, B. D., Tanielian, T. L., Stoto, M., & Lurie, N. (2005). In their own words: Lessons learned from those exposed to anthrax. *American Journal of Public Health, 95* (3), 489–495.

Brondolo, E., Wellington, R., Brady, N., Libby, D., & Brondolo, T. J. (2008). Mechanism and strategies for preventing post-traumatic stress disorder in forensic workers responding to mass fatality incidents. *Journal of Forensic and Legal Medicine, 15*(2), 78–88.

Brooks, S. K., Dunn, R., Amlôt, R., Greenberg, N., & Rubin, G. J. (2018). Training and post-disaster interventions for the psychological impacts on disaster-exposed employees: A systematic review. *Journal of Mental Health,* 1–25.

Castillo, C. (2004). Disaster preparedness and business continuity planning at Boeing: An integrated model. *Journal of Facilities Management, 31,* 8–26.

CDC (2020). *1918 Pandemic (H1N1 virus).* (2020). Available at: https://www.cdc.gov/flu/pandemic-resources/1918-pandemic-h1n1.html. Accessed 9 October 2020.

De Silva, D. A. M., & Yamao, M. (2007). Effects of the tsunami on fisheries and coastal livelihood: A case study of tsunami-ravaged southern Sri Lanka. *Disasters, 31*(4), 386–404.

Eyre, A. (1999). In remembrance: Post-disaster rituals and symbols. *The Australian Journal of Emergency Management, 14*(3), 23.

Eyre, A. (2006). Remembering: Community commemoration after disaster. In H. Rodríguez, E. L. Quarantelli, & R. R. Dynes (Eds.), *Handbook of disaster research* (pp. 441–455). NY: Springer.

Forrest, T. (1993). Disaster anniversary: A social reconstruction of time. *Sociological Inquiry, 63*(4), 444–456.

French, P. E., Goodman, D., & Stanley, R. E. (2008). Two years later: Hurricane Katrina still poses significant human resource problems for local governments. *Public Personnel Management, 37* (1), 67–75.

Goodman, D., & Mann, S. (2008). Managing public human resources following catastrophic events: Mississippi's local governments' experiences post—Hurricane Katrina. *Review of Public Personnel Administration, 28*(1), 3–19.

Hall, C. M., Malinen, S., Vosslamber, R., & Wordsworth, R. (Eds.). (2016). *Business and post-disaster management: Business, organisational and consumer resilience and the Christchurch earthquakes.* Routledge.

Henry, M., Cho, P., & Dupuis, P. (2008). Human resource development and personnel in a post-Katrina/Rita environment. *Community College Journal of Research and Practice, 32*(3), 220–234.

Ladika, S. (2006). Surviving Katrina-Remotely: Flexible, employee-centered actions by HR enabled a New Orleans hospital to stay open during and after the hurricane. *HR Magazine, 51*(1), 72.

Lambert, J., Duhon, D., & Peyrefitte, J. (2012). 2010 BP oil spill and the systemic construct of the Gulf Coast shrimp supply chain. *Systemic Practice and Action Research, 25*(3), 223–240.

Lippmann, M., Cohen, M. D., & Chen, L. C. (2015). Health effects of World Trade Center (WTC) Dust: An unprecedented disaster with inadequate risk management. *Critical Reviews in Toxicology, 45*(6), 492–530.

Malinen, S., Hatton, T., Naswall, K., & Kuntz, J. (2019). Strategies to enhance employee well-being and organisational performance in a postcrisis environment: A case study. *Journal of Contingencies & Crisis Management, 27*(1), 79–86.

Mankin, L. D., & Perry, R. W. (2004). Commentary: Terrorism challenges for human resource management. *Review of Public Personnel Administration, 24*(1), 3–17.

Mann, S. C., & Islam, T. (2015). The roles and involvement of local government human resource professionals in coastal cities emergency planning. *Journal of Homeland Security and Emergency Management, 12*(1), 145–168.

Massarweh, L. J. (2019). California burning: Staffing lessons learned from a multi-hospital system. *Nursing Economics, 37*(5), 241–245.

Nilakant, V., Walker, B., Rochford, K., & Van Heugten, K. (2013). Leading in a post-disaster setting: Guidance for human resource practitioners. *New Zealand Journal of Employment Relations, 38*(1), 1.

Norris, F. H., Friedman, M. J., & Watson, P. J. (2002). 60,000 Disaster victims speak: Part II. Summary and implications of the disaster mental health research. *Psychiatry, 65*(3), 240–260.

Norris, F. H., Friedman, M. J., Watson, P. J., Byrne, C. M., Diaz, E., & Kaniasty, K. (2002). 60,000 Disaster victims speak: Part I. An empirical review of the empirical literature, 1981-2001. *Psychiatry, 65*, 207–239.

Oxfam. (n.d.). *In the wake of the tsunami.* Available at: file:///C:/Users/brend/Desktop/BCP%20FINAL%20CHAPTER%20EDITS/oxfam-international-tsunami-evaluation-summary_3.pdf (Last Accessed May 29, 2020).

Paton, D. (1999). Disaster business continuity: Promoting staff capability. *Disaster Prevention and Management, 8*(2), 127–133.

Pennebaker, J. (1990). *Opening up: The healing power of confiding in others.* New York: Guilford Press.

Phillips, B. (2013). *Mennonite disaster service: Building a therapeutic community after the gulf coast storms.* Lanham, MD: Lexington Books.

Phillips, B. D. (2015). *Disaster recovery.* Boca Raton, FL: CRC Press.

Premeaux, S. F., & Breaux, D. (2007). Crisis management of human resources: Lessons from Hurricanes Katrina and Rita. *People and Strategy, 30*(3), 39.

Richardson, J. F. (2010). Disasters and remembrance: A journey to a new place. *Grief Matters: The Australian Journal of Grief and Bereavement, 13*(2), 49–52.

Srivastava, N., & Shaw, R. (2015). Occupational resilience to floods across the urban–rural domain in Greater Ahmedabad, India. *International Journal of Disaster Risk Reduction, 12*, 81–92.

Uusimäki, L. (2020). Beating stress, the Swedish way: Time for a 'Fika'. In L. McKay, G. Barton, S. Garvis, & V. Sappa (Eds.), *Arts-based research, resilience and well-being across the lifespan.* Cham: Palgrave Macmillan.

Wasileski, G., Rodríguez, H., & Diaz, W. (2011). Business closure and relocation: A comparative analysis of the Loma Prieta earthquake and Hurricane Andrew. *Disasters, 35*(1), 102–129.

Webb, G. R., Tierney, K. J., & Dahlhamer, J. M. (2002). Predicting long-term business recovery from disaster: A comparison of the Loma Prieta earthquake and Hurricane Andrew. *Global Environmental Change, Part B: Environmental Hazards, 4*(2), 45–58.

Woodward, N. (2006). Entry-level Gulf Coast workers are elusive. *HR Magazine, 26*, 34.

Yonder, A., Akcar, S., & Gopalan, P. (2005). *Women's participation in disaster relief and recovery*. The Population Council.

CHAPTER 7

Strengthening and testing your business continuity plan

Introduction

Plans alone are not enough. Too often, plans "sit on shelves collecting dust and giving a false sense of security or preparedness to their writers" (Perry, 2004, p. 65). Clarke (1999) refers to shelved plans as *fantasy documents* on which people rely to think that they are safe. It is not enough. People need to know how to enact the plan, which is best accomplished through training and exercise. Training therefore represents a crucial next step, where critical functions and work teams acquire the knowledge and skills necessary to implement the plan successfully. Exercises then evaluate both the performance of trained actions and the plan itself. Training and exercise must *both* be undertaken to ensure that the plan offers an effective and efficacious way to establish and maintain readiness. Do not let your plan get dusty on the shelf.

This chapter starts with an overview of why we train and how training might look for various enterprises. A section on exercises then outlines multiple options for conducting tests of the training and the plan. The section that concludes the chapter emphasizes creating and spreading a culture of safety and preparedness throughout an enterprise.

Training on the BCP

The purpose of this section is to inspire planning teams to train critical function teams, workplace leaders and supervisors, and others in positions of responsibility during simulated plan activation. Training centers on providing knowledge, skills, and abilities so that employees can maintain critical functions that promote business survival (Perry, 2004; Sinclair, Doyle, Johnston, & Paton, 2012). When a disaster happens, employees and workplace leaders will need to pivot quickly into their assigned roles. To do so, employees must know the plan to enact the plan. If they wait until a disaster to pull the plan out, read it, think about it, and then try to implement they will already be behind in

Business Continuity Planning. https://doi.org/10.1016/B978-0-12-813844-1.00001-4

taking critical actions. Line employees, supervisors, and managers must know and train to implement their work team assignments. Supervisors and managers must know how to lead through the emergency period to ensure successful plan implementation. Together, they can engage in actions that safeguard the business and their livelihoods. They are in this together.

Being ready to act thus serves as the key operational capacity required in a workplace. Accordingly, this section lays out rationales for why employees need to know the plan and offers options for businesses to organize training and exercises.

Why you need to know the plan

The Deepwater Horizon oil spill in the Gulf of Mexico in 2010 resulted in significant environmental impacts, lawsuits, health concerns, and investigations. During the inquiries, it became clear that the oil spill response plan had been written using what is called a "boilerplate" meaning that a templated plan had been copied and pasted from another context (Birkland & DeYoung, 2011). The plan included directions on how to rescue walruses, which do not live in the Gulf of Mexico. Credibility of the company dipped, and criticisms increased coupled with widespread negative media coverage and congressional inquiries. A company's reputation is at stake when the plan does not result in the best actions, not to mention the viability of the enterprise itself. These costs can far exceed the expenses related to developing real preparedness achieved through planning, training, and exercise.

As experienced researchers, your authors have also spent time talking to elected officials, emergency managers, business owners, agency heads and others who have had to deal with a disaster. It is clear, from decades of our own research, that if people do know their plan, they will not have time to pull it off the shelf and read it while the dust clears, the floodwaters recede, and the aftershocks stop (Perry, 2004). People need to lean into the plan well in advance of an event because of the limited time available when difficult things happen. Doing so will increase the chance of the business surviving, decrease potential costs, reduce anticipated losses, and avoid injuries. Training may also enable companies to address concerns about potential liabilities and lawsuits that arise from negligence around planning and preparedness. Training for such outcomes starts with using teachable moments to alert work teams to becoming more ready to act.

Teachable moments

In 2020, it felt like many of us took a collective step back and held our breath during the COVID-19 pandemic. As we waited to see how the pandemic would play out, which will do so after the publication of this volume, a teachable moment emerged in which we could look at how well prepared we were to weather this viral storm. The pandemic provided this moment in which many people could learn about the value of business continuity planning and our shared, collective responsibility for each other. Something similar happened in 1993, after the bombing of the World Trade Center claimed six lives and injured many people (Aguirre, Wenger, & Vigo, 1998; Wenger, Aguirre, & Vigo, 1994). In response to that event, a significant amount of preparedness, response, and continuity planning took place. Given that the World Trade Center buildings collapsed, with people trying to exit down stairwells in buildings with over 90 floors, it seemed to those who watched it happen that most working inside would have perished. Yet because of the planning, thousands of people survived by exiting the building far more rapidly than in 1993. One of the great stories of preparedness, training, and exercises leading to saved lives on 9/11 comes from the lesson implemented from the 1993 bombing by a financial services industry corporation, Morgan Stanley. The preparedness program and regular evacuation exercises in response to the 1993 bombing reduced potentially heavy losses. Although seven employees died, nearly 2700 others survived. Among them was the chief architect of the corporate preparedness initiative, decorated Vietnam Veteran and Vice President for Security, Rick Rescorla (Coutu, 2002). Clearly, training, exercising, and knowing what to do saves lives and enables businesses to continue their operations at some point.

Disasters that happen elsewhere, including those from the past, represent teachable moments when planners can secure the attention of employees, business executives, and supervisors. Taking advantage of a recent disaster allows the planning team to alert those around them. The evening news frequently includes reports of disasters that can be reviewed for relevance to the planning team's business. Details can be brought to the attention of employees at regular meetings, annual gatherings, or as part of a unit's goal-setting efforts. By integrating disaster readiness into regular meetings and events, planning becomes a more routine action for individuals and units to take. There is also value in creating specific events to raise disaster awareness and to inspire participation such as an annual preparedness day around a local hazard of concern. The onset of tornado or hurricane season, annual influenza periods, seasonal rains and flooding, or severe winter weather onset provide times when people may be more interested in and concerned with disasters. Seize the moment.

Training also makes employees feel more comfortable in their roles, because they come to know the various policies, procedures, and necessary actions. Following the Christchurch earthquakes in New Zealand, one study found that training and testing the plan has proven essential so that employees knew what to do (Hatton, Grimshaw, Vargo, & Seville, 2018). People also enjoy learning about disasters and how they can be more proactive in supporting the business that provides their paycheck (Alexander, Bandiera, & Mazurik, 2005). Training often results in employees taking information home to their families so that they are better prepared at home. Those who do are more ready to step in to their work roles with peace of mind about their families.

Training can be accomplished at reasonably low cost, by using existing facilities like conference rooms or classrooms (Idrose, Adnan, Villa, & Abdullah, 2007). Planners can also work with supervisors to create basic training scenarios or bring in consultants or trainers, which will increase costs but may provide useful expertise. More expensive options, such as simulations, can also be undertaken like Exercise Northstar that prepared Singapore hotels for a terror attack in the aftermath of the 2008 Mumbai attacks (Wee, 2017).

Training options

Training "is the activity that translates information defined as needed by the plan into a coherent program that can be imparted to responders" (Perry, 2004, p. 66). As an activity, training can be provided in multiple forms and should typically identify guidance and policies relevant to the action so that people can be informed. The business continuity planning team represents the best group to identify needed workplace training. They will know the organizational culture in which the training will occur, how employees may respond to new challenges, and the best method to deliver information to employees (Sinclair et al., 2012). The BCP team will also hold valuable insights to the critical functions and assigned work teams and be able to suggest or create related training.

Opportunities may also be available to leverage community partnerships to deliver low or no cost training to employees. For example, many law enforcement agencies deliver cost free active shooter preparedness training programs, such as the Citizen Response to Active Shooter Events (CRASE) program sponsored by Texas State University. Local Fire and Health Departments also deliver no cost Stop the Bleed training, sponsored by the Department of Homeland Security. Many other training opportunities may arise for specific hazards from other community partners. While training like CRASE and Stop the Bleed provide general preparedness skills for emergencies, the plan must also move beyond emergencies to consider the post-impact time. Many health care agencies conduct annual exercises to address potential hazards. In Ohio, the Ross-Pike-Hocking Health Care Coalition did just that in 2018 with a focus on an ice storm. Their goal? To discuss and train on how they would continue providing health care services in both congregate care and home-bound settings throughout a 7-day period with a crippling power loss and travel challenges. Such training can help to create a team of people who come to know and trust each other through their emergency training and to surface areas where they can collaborate to improve their planning to date.

What to train on

Planners should focus work teams on training to meet critical functions as specified in the BCP. That training can be general, such as a shift to telecommuting or training that requires highly specialized knowledge. For example, some settings require that specific steps must be taken sequentially for the proper shutdown of a system such as a nuclear plant. Sequential activities will require specific hands-on drills that train to the plan and enhance work team familiarity with essential tasks. Or, a business that relies heavily on computer-driven work like an investment firm or a tax accountant might need to train for how to initiate telecommuting, access computer systems remotely, and safeguard client information if they lose their workplace due to a flood or a power outage. A university headed into fall semester registration might want to surface and prepare materials to use for virtual registration during a pandemic or even how to use paper backups during an extended power outage. Training should thus invite people who must perform specific tasks related to a critical function into practicing so that they learn their tasks well and can step ably into performing them in a crisis.

In addition to practicing the plan, critical function work teams should also practice workarounds for specific tasks because disasters will disrupt what you want to do, when you want to do it, and how you want to do it. Thus, while training should certainly focus on the desired actions, training should also introduce a few surprises along the way to get people to think outside of the planning box. Astronauts, for example, train repeatedly on mission critical functions so that what they do in space becomes a familiar routine. But surprises do occur, which happened when an accident occurred on Apollo 13 as its crew headed to the moon. While the crew monitored the onboard situation and responded to recommendations from NASA, the crew's backup pilot on earth went directly into a simulator to surface workarounds including how to power up systems without compromising limited battery power. If the batteries failed, the crew would be lost. To their credit, the workarounds they created brought the crew safely home. The mission was lost, but the astronauts survived, and NASA continued to send crews into space for decades. Backups should always be invited to participate in the training because disasters may occur when people are away from the office, and someone else will need to step in.

Value exists also with inviting a range of participants to the training both as active trainees and observers. Differing roles and positions within the organization will require different knowledge, skills, and abilities (Kim, 2014). Supervisors, for example, need to know how employees are responding and to refine procedures as needed so they can manage a crisis well. Those same employees need to know that their supervisors know what needs to be done as well as how and why it is done in a particular way. Outside observers may also bring value, such as someone from a similar business who can observe neutrally or even the local emergency manager. Thus, training should involve people up and down the organizational hierarchy and from outside the enterprise, to have multiple eyes on how well the plan unfolds and what can be done to improve after the training ends.

Planning teams should establish a regular training and exercise cycle to ensure that work teams perform essential actions when a disaster occurs. In areas with repetitive and seasonally expected hazards, that training should occur before the expected onset of the hurricane, tornado, flood or influenza season. Training typically works best when it engages employees in not only the necessary training skills but as it applies to a potential and realistic scenario. Training should always be evaluated so that lessons learned can be integrated into regular plan revisions. Those revisions should occur at least annually to keep the plan as current as possible.

Thus, the goals of training are multi-fold. Not only does training help people practice procedures to accomplish assigned tasks, training also fosters understanding of how a specific critical function task supports the larger enterprise and enables business continuity. Training offers the opportunity to build more effective work teams through the necessary dialogue, coordination, and collaboration that will surface during a training event. Use training to build understanding of the plan and to bolster work teams into a disaster-ready team ready for any event. An important part of that includes creating training scenarios that realistic fit what might happen and challenge trainees to learn.

Creating scenarios

Scenario planning, and scenario-based training, represent useful options to generate interesting and relevant training. Scenario planning, which involves participants in looking at possible futures (in this case, specific to a disaster), has been increasing in use since September 11, 2001 (Molitar, 2009; Wilkinson, 2009). Scenarios, and the related scenario-based training help workplaces to be better prepared, to test the plan, to challenge beliefs about how things will happen, and consequently to improve the plan (Hiltunen, 2009; Roubelat, 2009). Much national planning has been scenario-based. For example, the U.S. Federal Emergency Management Agency used 15 scenarios to outline most of the planning and preparedness architecture for a post September 11th environment (FEMA, 2009).

Businesses worldwide, as well as human resource professionals, use scenarios to walk participants through thinking how they would and should react to situations that may arise in the future. BCP planning teams, then, should either craft scenarios (or hire a consultant) that will challenge participants to rethink how they might react in specific circumstances and to reconsider their roles in a future that has been harmed by a disaster. The goals of scenario planning thus task participants with understanding disasters better to see potential realities that will challenge them (Moats, Chermack, & Dooley, 2008). The goals include using a scenario to enable people to rethink what they believed would happen, what a different outcome might look like, and their place in a work team addressing critical functions around an untoward event.

BCP planners should be deeply involved in crafting scenarios. Even if a consultant organizes the training, the planning team will have the insights and ideas about what critical functions require testing. The business's HIRA will have surfaced the most likely events that a business will face, which should probably be the ones used to contextualize a training scenario. Involving the BCP planning team in creating the scenario will make it more realistic and will move them closer to their goals of building work teams capable of caring for what might happen.

Indeed, creating a scenario for training resolves around a "what if" discussion as in "what if we lost our internet to ransomware during tax season" and the place of business is an accounting firm. Or, "what if a flood entered our business to a level of one foot of water" and compromised files, desks, computers, and essential resources including a production/assembly line, access to key services, or areas required to keep the business operational? Or, "what if a tornado damaged" a specific area of a business like one of the dairy barns, or a floor that housed investigative records for a law enforcement agency, or a wing of a congregate care facility? Scenario writing requires that a realistic situation be described so that those tasked in the identified critical functions work teams can act. Planning teams will also know their businesses best and need to participate actively in crafting scenarios.

What might a scenario need to involve? These elements form a basis for what to consider when writing a scenario (see Box 7.1 and Fig. 7.1):

- Identify objectives intended to be assessed and evaluated in an exercise. For example, if a critical function focuses on restoring power to keep a business operational then "restoration of power" becomes something that an exercise can test participants on for their knowledge, skills, and abilities. Similarly, if transitioning to virtual client services represents a planned activity in a pandemic, then an exercise would walk participants through how they would make that pivot. An exercise evaluation template created by the U.S. Department of Homeland Security may be helpful to

BOX 7.1 Writing a sample scenario for a tabletop exercise.

Planners writing scenarios can pull from their hazard identification and the list of sources for hazard information indicated in Chapter 3. Additional resources can be downloaded free from https://www.fema.gov/hseep which serves as a repository for the U.S. Department of Homeland Security Exercise and Evaluation Program materials (Last Accessed June 18, 2020). Exercise plan templates for a full field exercise can be found at https://preptoolkit.fema.gov/web/hseep-resources/design-and-development (Last Accessed June 18, 2020).

A scenario might unfold like this:

Five days before today. It is mid-June with the annual rains coming through as expected. However, a tropical disturbance has arisen offshore that may be of concern. Though the storm does not reach hurricane-level speeds, it does appear to be moving slowly enough that forecasters have raised concerns about local rain impacts that may be heavier than normal.

- What should you be doing in your unit at this time given the forecast?
- Does everyone know the plan? If not, now is a good time to review work team assignments for critical functions.

Two days before today. Meteorologists and the National Weather Service have continued to focus on Tropical Storm Charlotte with have indicated that our area may be in the path of the storm. The "cone of uncertainty" that shows the storm path has edged closer to our area by the day.

- What should you be doing in your unit at this time given the forecast?
- Who is responsible for monitoring the forecast and alerting work teams assigned to critical functions?
- What systems might be affected if you experience water intrusion?
- What vendors might you need if you experience water intrusion?

Today. Tropical Storm Charlotte has shifted to move directly over our area. Although our business is located inland, the slow-moving storm is expected to sit and spin, dropping significant amounts of rain in the area. Local emergency managers have pushed out warnings for potential flooding and encouraged businesses and residents to stay out of flooded underpasses and to be alert for rising river levels.

- What should you be doing in your unit at this time given the heavy rainfall and area concern for flooding in your area?
- Which critical functions might need to be addressed? How can you get ready to do so?
- What is the plan to evacuate the facility if needed? Does everyone have a go kit in place to do so?
- Which employees, if any, might need to remain to address critical functions? Will they be in any danger if they are asked to remain? Should they remain? At what point will your business need to shut down with water intrusion? What actions should be taken now to continue business operations remotely?

Two days later. Seventeen inches of rain have fallen, raising the Bellport River above flood levels. Waters have spilled over the embankments and into the central business district, with nearly a foot of water entering dozens of businesses.

- What should you be doing in your unit at this time given that 12 in. of rain has flooded your business? The water also brought in mud and debris, ruining the business's first floor.
- Who has created a go kit to work away from a potentially flooded workplace?
- What critical functions will need to be taken care of first?
- Who are the essential workers who will be needed to restore critical functions?

One week later. Our business will not be able to function at normal capacity until flooring, furniture, sheetrock, insulation, and damaged resources will be replaced.

- What should you be doing in your unit at this time given that you must relocate for at least four to 6 weeks?
- What resources will you need to do that work away from your workplace? Was the go kit sufficient?
- Who may experience difficulties doing that work and what are your workarounds?
- Assume that employees have water in their homes. Who will complete their work?
- Is there work that cannot be done? How will that be addressed? Any possible workarounds?

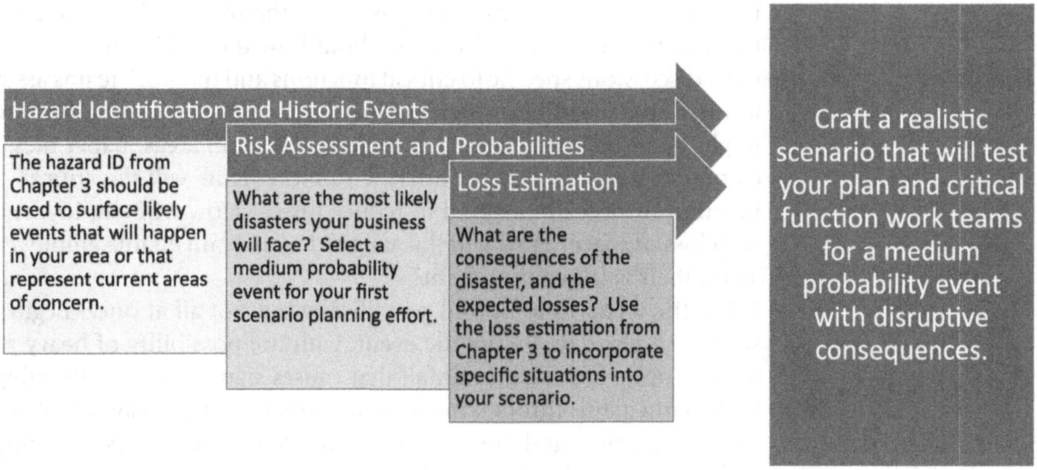

FIGURE 7.1
Writing a scenario.

surface training objectives and pull them from the critical functions in a BCP (https://preptoolkit.fema.gov/web/hseep-resources/eegs; see Economic Recovery download, Last Accessed June 18, 2020; see also the FEMA Business Continuity Planning Suite for a general template for exercise planning, https://www.ready.gov/business-continuity-planning-suite, Last Accessed June 18, 2020).

- Describe the potential future event using the hazard identification and hazards history developed out of planning team efforts in Chapter 3. Your research may have also surfaced photos or data that can make the training even more realistic.
- To determine which kind of scenario to craft, consider the analysis your team did in looking for the more commonly occurring event. Worldwide, floods represent the most common disaster, but your team might choose another more pressing scenario such as a cyberattack or pandemic. One way to determine which to choose is to look for frequently occurring events and to then pitch the scenario for a medium impact event. Low consequence events may not generate sufficient discussion while high consequence events may seem improbable to participants. Flooding that reaches one foot in height may be more realistic than one that reaches the second floor of a business – so aim for a scenario that provides for realistic discussion.
- Anticipate the disruptions (particularly downtime and displacement) that will occur from the scenario and integrate them into the scenario.
- Anticipate the kinds of losses that may occur, which should have been revealed in your prior loss estimation. Your planning team can use that

loss estimation to set up situations where the scenario-based training participants must make decisions about how they will cope.

- Craft discussions specific to critical functions and the work teams assigned to them. If a flood inundates the first floor of an accounting firm, what might be lost? First floors often include reception areas, paper files, computer supplies, power outlets, and more. How will the critical function work team respond to reduce losses? How will employees cope with loss of space? How will the assigned work team enable employees to adapt their work environment?

- Allow the scenario to unfold periodically but not all at once. Begin the scenario a few days before the event, with the possibility of heavy rains leading up to significant rainfall that causes significant area flooding. Walk participants through the time that follows with a heavy emphasis on the recovery time and how employees would cope with loss of workspace and resources to do their jobs. Bear in mind that some participants may be "new" to the scenario and business continuity planning process so they may be uncertain. Invite everyone to participate and tell them it is a learning environment where it is safe to explore what one would – and would not – want to do.

- Focus on "what will you do" at the various points in your scenario and ask participants to discuss their roles. A timeline helps with creating points when participants will pause and consider key questions through discussion.

- Gather lessons learned from the discussion to revisit the BCP and make adaptations – because planning teams always learn new information when they interact with the work teams and employees tasked with putting things back in order. Do not feel that you have forgotten something, because everyone brings a new perspective to the planning and training tables.

- Involve a wide set of stakeholders, from the assigned work teams, to the people likely to be affected (Keough & Shanahan, 2008). Involve everyone across various functional areas and levels to encourage differences of opinion and perspectives to emerge (Keough & Shanahan, 2008). There is nothing worse than having someone remain quiet when they might have a great idea or prevent a horrendous error from being made – remember that groupthink is very real, with the only way to overcome it to enable everyone's voices to be heard. A skilled facilitator will be required to make this happen (Chen, 2009).

- Ask participants afterwards to critique their actions, identify lessons learned, and surface actions that need to be taken. Doing so will then inform both the plan and the work teams assigned to specific critical functions. It is very important that planners encourage people to be open and honest, as few lessons are learned from people patting themselves on the back without deep and serious reflection.

- Organize your scenario into a script or PowerPoint that walks trainees through various decision points.
- Ideally, scenario-based training would include a pre- and post-training assessment to measure knowledge, skills, and abilities so that the plan - and the training protocol – can be modified as needed (Idrose et al., 2007; Sinclair et al., 2012).

Training represents only the first step in making sure a business continuity plan can be implemented. Testing the training requires a larger scale with more people. Within emergency management, that level of efforts is accomplished through various kinds of exercises.

Exercises

Training is considered a first step in being able to implement a plan and stand in contrast to exercises. Exercises provide carefully crafted opportunities to examine how well employees have acquired necessary knowledge, skills, and abilities to enact the plan (Perry, 2004; Sinclair et al., 2012). Following the Christchurch earthquakes, businesses realized that existing plans lacked sufficient detail – exercises can reveal those gaps before an event (Hatton et al., 2018). Exercises thus represent an operational test of the critical function work teams. Opportunities to engage in a full exercise lets participants get their hands on equipment, props, computers, and cleaning supplies and to surface creative ways to meet critical functions (Peterson & Perry, 1999).

Interestingly, exercises can provide additional benefits (Cotanda, Martínez, de la Maza, & Cubells, 2016). Exercises also allow those tasked with critical functions to develop working relationships with their fellow employees as they collaborate and problem-solve in a quasi-realistic event (Perry, 2004; Peterson & Perry, 1999). Exercises thus make plans more real and alert participants and the broader workforce to the value of preparedness (Perry, 2004). Social psychological results also accrue for participants, including enhanced teamwork, a sense of readiness, and confidence in knowing how to take specific actions to save the business (Perry, 2004). Exercises provide significant value to employees and workplaces and should be undertaken at some level on a regular basis.

Exercises should have some common elements. First, like with training, an exercise should set out specific objectives, such as participants' training for implementing a post-disaster business continuity planning protocol (Perry, 2004). Second, an exercise typically revolves around a scenario just like training, usually a narrative set up by a planner or skilled facilitator and linked to a locally expected threat (Perry, 2004; Savoia, Biddinger, et al., 2009; Savoia, Testa, et al., 2009). Blades (2015) advises that senior management should be involved in the exercise concept development, as should planners. Exercises should be realistic and challenge participants to think through their roles and to anticipate curveballs that might happen. Exercises should tell planners a great deal about the adequacy of their planning and represent "an avenue to continuous quality improvement" (Peterson & Perry, 1999, p. 254).

In terms of a format, exercises can vary from a seated, tabletop discussion to people acting out their roles in a disaster simulation with equipment, props, and other people (Perry, 2004). Businesses can choose an exercise format and each offers varying depths to test a plan and participants' knowledge, skills, and abilities (Sharpe, 2017) and at varying costs. The U.S. Department of Homeland Security (2020) maintains exercise and evaluation guidelines through the Homeland Security Exercise Evaluation Program (HSEEP) (see Box 7.1). From easiest and least expensive to the most complex and expensive, options include (in order):

Tabletop exercises (TTX). A tabletop exercise typically happens with participants listening to a scenario while gathered at a conference table or similar location. Tabletops can be straightforward and short in length or can involve participants in a day-long activity with skilled facilitators and breakout leaders. A facilitator leads participants verbally through what they expect to do (see Fig. 7.2) by having participants focus on their own roles in a potential disaster and then join in a general discussion. The TTX provides an opportunity to discuss how the plan seems to be working as well as what might have been neglected (Perry, 2004). As the most cost-effective exercise, a TTX gives businesses a chance to test a plan at a general level prior to organizing or paying for more expensive formats. Tabletops also get the plan out of people's imagination and into a shared discussion that can surface additional questions, concerns, issues, and solutions. Tabletops also offer a lower stress environment in which to first play out one's potential roles in disaster and can be less intimidating than a more in-depth format. Tabletops can focus on a specific critical function or incorporate teams across a business. Tabletops can also include community partners, vendors, and others important to a business's recovery. Careful design and involvement of all participants is important to promote critical thinking. A report by De Allicon (2020) reports the benefits of incorporating principles of mindfulness into BCP exercises to support mental presence and awareness in exercises so important to business continuity.

The development of tabletop exercises does not have to rely solely on the shoulders of a business continuity planner. In the U.S., many no-cost resources are available to support certain hazard event scenarios. The Department of

FIGURE 7.2
Participants join a tabletop exercise on cyber security between governmental and business partners. *Melissa Wiehenstroer/FEMA News Photo.*

Homeland Security (DHS) Critical Infrastructure Security Agency (CISA) provides resources to support tabletop exercises, known as a "tabletop in a box." For businesses in critical infrastructure sectors, materials can be secured through the DHS Homeland Security Infrastructure Network (HSIN) portal. The DHS Campus Resilience Program provides specific exercise starter kits for educational organizations for a variety of hazards that are tailored to the k-12 and higher education environments (https://www.dhs.gov/academicresilience). FEMA also provides tabletop support resources for active shooter event exercises (https://www.fema.gov/media-library/assets/documents/181757). The FEMA Private Sector Division maintains tabletop starter resources for critical power failures, hurricane, and chemical accidents (https://www.fema.gov/emergency-planning-exercises#). These resources provide strong starting points adaptable to almost any circumstance. However, an existing tabletop starter resource should always be redesigned to fit local hazards and business context.

Tabletops can also be as complex as more engaging exercises but will require a significant amount of resources and collaboration among those involved. This is particularly true of the complexities involved in health care businesses (Dausey, Buehler, & Lurie, 2007). For example, multiple, relatively short-lived pandemics that occurred in the last 30 years have generated a significant amount of pandemic planning and related templates. To make it realistic, the state of Maryland in the U.S. created a tabletop exercise involving potential actors who would enact the plan (Taylor et al., 2005). The tabletop exercise set out a series of assumptions for a pandemic to occur such as the way in which the disease would spread, the number of people that would become ill and die, and the timeline for a vaccine. The scenario included one main pandemic description with multiple outcomes and played out over several months with participants stopping to answer questions as needed. For example, one set of events and related discussion centered on schools being affected coupled with questions about when and how schools and universities would close. Does this sound familiar to those who experienced COVID-19 in 2020?

Some businesses or agencies might want to broaden their tabletop exercises beyond their own boundaries. A chemical accident, for example, may affect not only the physical plant where the accident happened but the surrounding community. Tabletop exercises can enable potential actors to come together, build relationships, and consider how they will interact in a specific situation. Formally coming together can also help to build a culture of preparedness that can further repel a disaster and ready those who will need to be involved in the response and recovery. One study of actors concerned with a chemical accident in a convention center found that over half of the participants improved their understanding of their roles (High et al., 2010). A tabletop exercise thus affords an excellent opportunity to bring in community response organizations and represents a low-cost commitment for emergency management, law enforcement, fire departments, public health, and/or public works officials to participate. They may be included later in a series of tabletop exercises that may follow initial iterations with facility staff and leadership only.

Functional exercises. A second kind of exercise can occur at the tabletop level or in the full field exercises described next. Functional exercises focus on a specific critical function. Typically, a functional exercise promotes very focused testing of staff training, knowledge, skills, and abilities. That could range from how to pivot from traditional restaurant service to takeout or from a reliance on utilities to setting up a generator. In short, a functional exercise usually centers on just a single goal or sub-goal to test a specific set of knowledge or part of a plan. One example of a limited functional exercise or a drill is testing lockdown procedures in the event of an active shooting event. An exercise for this function can include objectives related to the recognition of the hazard, communication of the situation, and the implementation of self-protection measures by staff, such as avoiding injury by leaving, denying entry to areas by locking and barricading doors, and preparing to defend oneself if other options are not available. A functional exercise such as this can familiarize employees with the facility, identify safety options, and surface maintenance issues and necessary facility upgrades. An important underlying issue for exercises of this nature are to ensure that all involved, local response organizations are aware of the exercise. Poorly managed communications can lead to injuries, embarrassment, and loss of trust in policy, plans, and organizational leadership. In one example, failures in communication led to an active shooter exercise on Wright Patterson Air Force Base in 2018 to be interpreted as an actual event, resulting in an overwhelming response. A responding law enforcement officer fired shots through a door to make entry that caused injury (Air Force Times, 2019). Appropriate management of all exercises is necessary, but caution must be used when any potential injury can occur. Communication represents one of the most critical skills that needs to be cultivated fully in exercises (Klima et al., 2012).

Full-scale exercises. A full-scale or field exercise takes place at a given location in the business or perhaps in concert with other businesses. Firefighters, police, and paramedics engage in full-scale exercises on a regular basis. Health care settings often participate, such as exercising a plan for a hazardous materials spill. For response organizations and healthcare settings these type of exercises are periodically required for accreditation purposes or to receive certain funding. For business continuity planning, the emphasis lies on what people will do after the disaster has happened to restore critical functions to the business operations. That might include getting financial planning systems and financial advisors back online by testing an offsite location or working from home. Educational systems might conduct an exercise by switching classes from face-to-face to online delivery for a pandemic or a blizzard.

A full-scale exercise will require a significant amount of advance work and resources as such efforts typically try to test multiple parts or the entire plan with many goals and sub-goals (Perry, 2004). A very careful scenario and script will need to be crafted. Depending on the nature of the exercise, extra role players, participants, facilitators, and evaluators may need to be identified and briefed such as when an airport conducts a crash exercise or a health care setting sets up a practice Point of Distribution (POD) for vaccines. Full-scale exercises necessitate coordination and collaboration well in advance of trying to undertake the initiative. The moving parts also require a high focus on safety. An exercise one of the authors was heavily involved in planning and executing involved a scenario with a criminal hazardous materials incident resulting in a mass casualty event on a train. The exercise included incident response objectives and business continuity objectives for the contract operator of the rail line. The planning took nearly 5 months with as many as 50 attendees at bi-weekly planning meetings. The exercised cost in-excess of $100,000 and involved moving a train to a station closed for the weekend, dozens of role players, and more than 50 responders.

For smaller businesses with less resources, a full field exercise might take place at a smaller scale. An assisted living facility, for example (where people tend to be more mobile than a nursing home), could practice moving residents to another location. Such a scenario might occur with a power loss or an approaching wildfire. Such a fire threatened 300 residents of New Jersey nursing homes in 2007 as did a wildfire in Oklahoma a few years later. Moving people to a convention center or recreational area where the residents can then enjoy an outing may enable both the practice and benefit the residents for a half-day event. But the full field exercise should not just focus on transportation – what else might be needed if the move were a real one that lasted several days? What might need to be at the convention center to receive the assisted living residents? Which external organizations should be involved in the exercise? What kinds of records, resources, and go kits might be needed? Critical functions established in the planning process should have identified what would be needed and who should do it. This kind of full-field exercise can also be done without residents and with actors like college students to provide a realistic exercise experience.

Virtual/simulations. A final option might come from a virtual simulation that allows participants to exercise their plan. Typically, simulations involve participants in working through a visualization of a scenario using computerized technology. Some facilities provide places to conduct a virtual simulation, like a decontamination station at a hospital parking lot or a nursing lab that emulates an emergency department (Alexander et al., 2005). The hotel industry in Singapore has worked at being ready for a terrorist attack as mentioned earlier in this chapter (Wee, 2017). Desk clerks, for example, train on how to record guest information and room assignments correctly and then back it up at an offsite location using computer simulations. Regular training and audits insure that employees are ready to assist and account for guests in an emergency.

While options certainly exist, acquiring them may prove expensive and require external contracts with specialized vendors. However, virtual simulations already exist for many professional places like emergency management agencies, hospitals and health care settings, and hospitality industries and may represent the future of exercises for businesses. Some simulations may prove more accessible and even more effective than the costs involved such as how to evacuate a large building or a cruise ship (Strohschneider & Gerdes, 2004). Virtual simulation that train passengers how to exit an airplane have resulted in better understanding and action than handing passengers a card marking exit procedures (Chittaro, 2016). Multiple opportunities thus exist to test the training and the plan through various kinds of exercises. Ideally, training and exercising will help to establish a level of readiness significantly higher than before that helps to reduce impacts.

Creating a culture of safety and preparedness

Training and exercises enable work teams to test their plans, learn more about their roles, and surface new ideas and insights. But training and exercises alone are not sufficient to ensure that a business will prove to be resilient while moving through and into the aftermath of a disaster. To prove truly able to resist the impacts of a disaster and to rebound from disruption that threatens the bottom line, the business's mission, and employee paychecks requires creating an organizational culture of safety and preparedness (Kim, Park, & Park, 2016). Defined, a culture represents a way of life or a design for living that includes behavioral norms based on shared values. An *organizational or business culture* lays out values that influence how we behave with and toward each other in a way that promotes safety and increases business viability beyond an untoward event. A culture of safety and preparedness emphasizes the behaviors that lead to personal and organizational *readiness to enact emergency and business continuity plans*.

For organizations where the actual work presents substantial risks to employees, such as manufacturing, much has been written about developing a safety culture (see DeJoy, 2005; Fernández-Muñiz, Montes-Peón, & Vázquez-Ordás, 2007). In the U.S., occupational safety is driven by the Occupational Safety and Health Administration (OSHA). In Chapter 5, we noted how OSHA requires employers with a work site that has more than 10 employees to have a written Emergency Action Plan with training requirements. This and other employee protection requirements can drive a focus on employee safety which can be the foundation for enjoining preparedness with safety. A robust and effective workplace safety program results in fewer employee injury claims, decreased loss of productivity, and lower employee injury claims that leads to lower insurance costs.

Further, good management practices decrease work injury rates (Ali, Abdullah, & Subramaniam, 2007). Strong leadership, role modeling, and supervision drive employee behavior for both safety (Fernández-Muñiz et al., 2007) and preparedness. Famed industrial organizational psychologist Edgar Schein advises that the leader influences culture through five mechanisms (Schein, 1990): (1) What leaders pay attention to, measure and control; (2) How leaders react to critical incidents and organizational crises; (3) Deliberate role modeling and coaching; (4) Operational criteria for the allocation of rewards and status; and (5) Operational criteria for recruitment, selection, promotion, retirement, and excommunication (Schein, 1990, p. 115). In short, people follow leaders who exhibit behaviors that compel employees to follow planning around emergency and business continuity goals.

Creating a culture of preparedness means that employees will be more ready for the primary hazard that the business continuity planning team identified. An example comes from recent threats that have caused computer systems to fail and entire enterprises – banks, hospitals, and universities – to lose valuable time, face the costs of ransomware, and risk financial impacts due to lawsuits. In response, workplaces have launched efforts to enhance employee understanding of information security (Tang & Zhang, 2016). Employees today can routinely expect to receive notices about phishing, where thieves attempt to secure usernames, passwords, and other information to access bank accounts, payroll, personal information, and more (Arachchilage, Love, & Beznosov, 2016). Workplaces may even create internal phishing look-alikes to raise awareness with their employees. By raising the alert level of employees working through email in their inboxes, cybersecurity will increase, and threats will decrease. Efforts are designed to result in behavioral changes that improve resistance to cyber intrusions.

Efforts to influence behavior must thus be focused within an organization on the hazards most likely to present themselves, which planners surfaced in their risk assessment (see Chapter 3). Knowing what hazards must be addressed, the BCP team or designated employees can design initiatives to influence employee behavior. The company's IT team or consultant can be asked to design educational initiatives that raise awareness about cyberthreats. A health insurance company or an internal health unit can promote behaviors that encourage handwashing or getting a flu shot to reduce employee absences – or a pandemic. Everyone needs to be weather-alert during severe weather season. Regardless of the threat, many steps can be taken to promote preparedness and safety.

For example, natural hazards and influenza have seasons when they routinely occur. Tornado season in the U.S. and Canada starts March 1, which presents an opportune time to raise awareness about personal risks and how a workplace will issue alerts, deal with a direct strike, and emerge ready to continue operations. Likewise, Australia's bushfire season typically peaks in January and February during the heat of their summer. Businesses can educate employees about risks, response, and recovery planning prior to the onset of the season. They can involve employees in risk reduction efforts as well from installing fire-resistant landscaping to how and where a business will relocate if bushfire impacts their site. The pre-season time of bushfire can thus alert people to risk and what to do (Whittaker, Taylor, & Bearman, 2020).

Businesses may also take advantage of specific days to educate and alert employees to various hazards. Many locations or nations specify a time to raise awareness. New Zealand does that, with a national earthquake "shakeout" day, an effort that began in the U.S. (Becker et al., 2016; Jones, 2020). The purpose of the drill is to have a designated day and time when people must take protective actions such as stop, drop, and cover yourself under a desk. The shakeouts could also have an additional element, where BCP teams then ask employees to describe what they would have done if the building had been damaged to the point where they could not complete their work for several days or weeks – that would help to inform employees across a business in addition to the specific members of the BCP work teams.

Employers can also create an annual Safety Day within the business to recognize employees who have addressed safety matters. While the focus of this volume is on disasters, recognizing a range of hazards can serve to increase awareness about safety, disasters, and catastrophes. Many businesses proudly erect signs indicating the number of injury-free days on the job. While injuries to individuals are far different from disasters that destroy buildings, by emphasizing individual safety – and the responsibility of employees for each other – employers can create an organizational culture that embraces safety and preparedness – and as a foundation for post-disaster resilience. Business continuity planners and their work teams certainly deserve recognition as do employees who earn certificates for safety, emergency management, cybersecurity, and other relevant forms of training. In the U.S., the Department of Homeland Security sponsors September as National Preparedness Month (see preparedness materials at https://www.ready.gov/september) and October as National Cybersecurity Awareness Month (see Cybersecurity materials at https://www.cisa.gov/national-cyber-security-awareness-month). DHS provides communication materials and suggested activities for each week of the month that can be incorporated into employee newsletters, social media, and employee programs. Engaging these programs can help focus on preparedness and cybersecurity issues reinforcing importance of these issues to the organization.

Various homeland security agencies also use threat systems to alert the public and potentially affected areas to an increased risk. By monitoring various forms of communication, analysts can determine if a potential attack is imminent. Those warning levels can help to catch the attention of businesses in high-target, high-traffic areas like malls, subways, and larger gatherings. The most critical recent communication to business operators came in the wake of several vehicle-ramming attacks on crowded pedestrian venues in several countries in 2016 and 2017. The U.S. Department of Homeland Security (2017) distributed security awareness materials targeted at operators of potential targets providing awareness of attack indicators, emergency response actions, and mitigation and protective action strategies. For a broader array of threats, many business sectors in the U.S. collaborate through an Information Sharing and Analysis Center (ISACs) that conducts specific research, analysis, and notification for threats that may impact businesses in the sector. Twenty-five ISACs ranging from the American Chemistry Council to the Water ISAC provide tailored threat information to industry partners (for a full list and information on joining an ISAC see the National Council of ISACs – https://www.nationalisacs.org/).

To conclude, consider the 2020 COVID-19 pandemic. Taiwan has been building a culture of safety since the SARS outbreak in 2003, which was followed by concerns with H5N1 and H1N1. Taiwan's strategies have included a seasonal, mass vaccination program along with now-familiar strategies including social distancing, handwashing, and mask-wearing with an emphasis on teaching children about their health-related responsibilities (Meyer et al., 2018; Schwartz & Yen, 2017). By mid-June of 2020, Taiwan had reported seven deaths to COVID-19 while the United States continued climbing past 130,000 lost lives. The COVID-19 outbreak resulted in many organizations emphasizing health and safety to thwart a second or third wave of the deadly pandemic in what may become a new way of life. Grocery stores placed Plexiglas barriers between checkout clerks and customers. Contactless payment emerged to prevent contamination through handling cash or debit and credit cards. Wearing masks in public also became a norm in many countries, businesses, and for airport travelers as well as people enjoying a social time with friends. An emphasis on handwashing, using specific techniques, generated posters, videos, songs, and memes to spread the importance of personal responsibility for others. If we truly want to reduce the impacts of disasters, we must cultivate further a culture of preparedness that enhances personal and workplace safety all the time, for all hazards, and all employees. Then, businesses will stand the best chance of survival as part of their business continuity efforts.

Essential actions

This chapter addressed the importance of education and training around the business continuity plan. BCP teams should therefore:

- Organize efforts to train work teams on the business continuity plan and their roles should a disaster occur.
- Outline scenarios that can be used to alert work teams and business employees to what might happen, including the critical functions that must be addressed to reduce expected losses.
- As part of business continuity preparedness programs offer realistic exercises, whether they are tabletop events or full-scale exercises that involve the entire business, their partners, and the affected community.
- Develop multiple ways to foster an organizational culture that values safety and preparedness. Keep disaster awareness on people's minds appropriately from workplace drills to an annual safety day. Participate in

broader campaigns such as Preparedness and Cybersecurity awareness months.

- Recognize employees who support and demonstrate safety and preparedness.
- Update and revise business continuity plans at regular interviews, at least annually. Be sure that contact lists remain current, that critical functions continue to reflect the core missions of the business, and that work teams can still step up to fulfill those functions.

References

Aguirre, B. E., Wenger, D., & Vigo, G. (June 1998). A test of the emergent norm theory of collective behavior. *Sociological forum* (pp. 301–320). Vol. 13(2)(pp. 301–320). Kluwer Academic Publishers-Plenum Publishers.

Air Force Times (August 1, 2019). *Wright-Patt makes changes after 'active shooter' reaction. Retrieved from (August 1, 2019). https://www.airforcetimes.com/news/your-air-force/2019/08/01/wright-patt-makes-changes-after-active-shooter-reaction/.*

Alexander, A. J., Bandiera, G. W., & Mazurik, L. (2005). A multiphase disaster training exercise for emergency medicine residents: Opportunity knocks. *Academic Emergency Medicine, 12*(5), 404–409.

Ali, H., Abdullah, N., & Subramaniam, C. (2007). Management practice in safety culture and its influence on workplace injury: An industrial study in Malaysia. *Disaster Prevention and Management, 18*(5), 470–477.

Arachchilage, N. A. G., Love, S., & Beznosov, K. (2016). Phishing threat avoidance behaviour: An empirical investigation. *Computers in Human Behavior, 60,* 185–197.

Becker, J. S., Coomer, M. A., Potter, S. H., McBride, S. K., Lambie, E. S., Johnston, D. M., & Walker, A. (April 2016). Evaluating New Zealand's "ShakeOut" national earthquake drills: A comparative analysis of the 2012 and 2015 events. In: *Proceedings of the 2016 NZSEE conference, Christchurch, New Zealand,* pp. 1–3.

Birkland, T. A., & DeYoung, S. E. (2011). Emergency response, doctrinal confusion, and federalism in the Deepwater Horizon oil spill. *Publius: The Journal of Federalism, 41*(3), 471–493.

Blades, A. (2015). Business continuity exercises: Concept development the foundation for success. *Governance Directions, 67*(2), 93–95.

Chen, J. K. (2009). Utility and drawbacks of scenario planning in Taiwan and China. *Journal of Futures Studies, 13*(3), 105–106.

Chittaro, L. (2016). Designing serious games for safety education: "Learn to brace" versus traditional pictorials for aircraft passengers. *IEEE Transactions on Visualization and Computer Graphics, 22*(5), 1527–1539.

Clarke, L. (1999). *Mission improbable: Using fantasy documents to tame disaster.* Chicago: University of Chicago Press.

Cotanda, C. P., Martínez, M. R., de la Maza, V. T. S., & Cubells, C. L. (2016). Impact of a disaster preparedness training programme on health staff. *Anales de Pediatría (English Edition), 85*(3), 149–154.

Coutu, D. L. (2002). How resilience works. *Harvard business review on point,.* Retrieved from (2002). https://hbr.org/2002/05/how-resilience-works.

Dausey, D. J., Buehler, J. W., & Lurie, N. (2007). Designing and conducting tabletop exercises to assess public health preparedness for manmade and naturally occurring biological threats. *BMC Public Health, 7*(1), 92.

De Allicon, K. (2020). A mindfulness toolkit to optimise incident management and business continuity exercises. *Journal of Business Continuity & Emergency Planning, 13*(3), 220–229.

DeJoy, D. M. (2005). Behavior change versus culture change: Divergent approaches to managing workplace safety. *Safety Science, 43*(2), 105–129.

Federal Emergency Management Agency (2009). *National planning scenarios fact sheet. Retrieved from (2009). https://www.fema.gov/txt/media/factsheets/2009/npd_natl_plan_scenario.txt.*

Fernández-Muñiz, B., Montes-Peón, J., & Vázquez-Ordás, C. (2007). Safety culture: Analysis of the causal relationships between its key dimensions. *Journal of Safety Research, 38*, 627–641.

Hatton, T., Grimshaw, E., Vargo, J., & Seville, E. (2018). Lessons from disaster: Creating a business continuity plan that really works. *Journal of Business Continuity & Emergency Planning, 10*(1), 84–92.

High, E. H., Lovelace, K. A., Gansneder, B. M., Strack, R. W., Callahan, B., & Benson, P. (2010). Promoting community preparedness: Lessons learned from the implementation of a chemical disaster tabletop exercise. *Health Promotion Practice, 11*(3), 310–319.

Hiltunen, E. (2009). Scenarios: Process and outcome. *Journal of Futures Studies, 13*(3), 151–152.

Idrose, A. M., Adnan, W. A. W., Villa, G. F., & Abdullah, A. H. A. (2007). The use of classroom training and simulation in the training of medical responders for airport disaster. *Emergency Medicine Journal, 24*(1), 7–11.

Jones, L. M. (2020). Empowering the public with earthquake science. *Nature Reviews Earth & Environment, 1*(1), 2–3.

Keough, S. M., & Shanahan, K. J. (2008). Scenario planning: Toward a more complete model for practice. *Advances in Developing Human Resources, 10*(2), 166–178.

Kim, H. (2014). Learning from UK disaster exercises: Policy implications for effective emergency preparedness. *Disasters, 38*(4), 846–857.

Kim, Y., Park, J., & Park, M. (2016). Creating a culture of prevention in occupational safety and health practice. *Safety and Health at Work, 7*(2), 89–96.

Klima, D. A., Seiler, S. H., Peterson, J. B., Christmas, A. B., Green, J. M., Fleming, G., ... Sing, R. F. (2012). Full-scale regional exercises: Closing the gaps in disaster preparedness. *Journal of Trauma and Acute Care Surgery, 73*(3), 592–598.

Meyer, D., Shearer, M. P., Chih, Y. C., Hsu, Y. C., Lin, Y. C., & Nuzzo, J. B. (2018). Taiwan's annual seasonal influenza mass vaccination program—Lessons for pandemic planning. *American Journal of Public Health, 108*(S3), S188–S193.

Moats, J. B., Chermack, T. J., & Dooley, L. M. (2008). Using scenarios to develop crisis managers: Applications of scenario planning and scenario-based training. *Advances in Developing Human Resources, 10*(3), 397–424.

Molitar, G. T. (2009). Scenarios: Worth the effort. *Journal of Futures Studies, 13*(3), 81–92.

Perry, R. W. (2004). Disaster exercise outcomes for professional emergency personnel and citizen volunteers. *Journal of Contingencies & Crisis Management, 12*(2), 64–75.

Peterson, D. M., & Perry, R. W. (1999). The impacts of disaster exercises on participants. *International Journal of Disaster Prevention and Management, 8*(4), 241–254.

Roubelat, F. (2009). Scenarios in action: Comments and new directions. *Journal of Futures Studies, 13*(3), 93–98.

Savoia, E., Biddinger, P. D., Fox, P., Levin, D. E., Stone, L., & Stoto, M. A. (2009). Impact of tabletop exercises on participants' knowledge of and confidence in legal authorities for infectious disease emergencies. *Disaster Medicine and Public Health Preparedness, 3*(2), 104–110.

Savoia, E., Testa, M. A., Biddinger, P. D., Cadigan, R. O., Koh, H., Campbell, P., & Stoto, M. A. (2009). Assessing public health capabilities during emergency preparedness tabletop exercises: Reliability and validity of a measurement tool. *Public Health Reports, 124*(1), 138–148.

Schein, E. H. (1990). Organizational culture. *American Psychologist, 45*(2), 109–119.

Schwartz, J., & Yen, M. Y. (2017). Toward a collaborative model of pandemic preparedness and response: Taiwan's changing approach to pandemics. *Journal of Microbiology, Immunology and Infection, 50*(2), 125–132.

Sharpe, C. (2017). Plan of action: A library's journey to training for emergencies and disasters. *Journal of New Librarianship, 2*, i.

Sinclair, H., Doyle, E. E., Johnston, D. M., & Paton, D. (2012). Assessing emergency management training and exercises. *Disaster Prevention and Management, 21*(4), 507–521.

Strohschneider, S., & Gerdes, J. (2004). MS ANTWERPEN: Emergency management training for low-risk environments. *Simulation & Gaming, 35*(3), 394–413.

Tang, M., & Zhang, T. (2016). The impacts of organizational culture on information security culture: A case study. *Information Technology and Management, 17*(2), 179–186.

Taylor, J. L., Roup, B. J., Blythe, D., Reed, G. K., Tate, T. A., & Moore, K. A. (2005). Pandemic influenza preparedness in Maryland: Improving readiness through a tabletop exercise. *Biosecurity and Bioterrorism: Biodefense Strategy, Practice, and Science, 3*(1), 61–69.

U.S. Department of Homeland Security (2017). *Vehicle ramming security awareness for soft targets and crowded places. Retrieved from(2017).* https://www.cisa.gov/sites/default/files/publications/Vehicle%20Ramming%20-%20Security%20Awareness%20for%20ST-CP.PDF.

U.S. Department of Homeland Security (2020). *Homeland security exercise and evaluation program (HSEEP). Retrieved from(2020).* https://www.fema.gov/media-library-data/158266986265094efb02c8373e28cadf57413ef293ac6/Homeland-Security-Exercise-and-Evaluation Program-Doctrine-2020-Revision-2-2-25.pdf.

Wee, K. T. H. (2017). Exploring a new approach to business continuity management training practices by Singapore hotels to manage terrorist threats. *Asian Journal of Public Affairs, 17*–22.

Wenger, D., Aguirre, B., & Vigo, G. (1994). *Evacuation behavior among tenants of the World Trade Center following the bombing of February 26*: (p. 1993). Hazards Reduction and Recovery Center: College Station, TX.

Whittaker, J., Taylor, M., & Bearman, C. (2020). Why don't bushfire warnings work as intended? Responses to official warnings during bushfires in New South Wales, Australia. *International Journal of Disaster Risk Reduction,*101476.

Wilkinson, A. (2009). Scenarios practices: In search of theory. *Journal of Futures Studies, 13*(3), 107–114.

CHAPTER 8

Becoming more resilient

Introduction

Congratulations on reaching the final chapter in the book and the end of the formal planning process, with the exception for regular training and exercising, and annual updates. Your planning team will have created a core business continuity plan with work teams dedicated to restoring critical functions when crisis occurs. Training and exercising on the plan will have brought new issues to light, hopefully resulting in updates and new ideas being integrated into plan revisions. Training and exercising will have also created a team alert to a potential disaster that will look ahead, see something coming, and implement the plan.

Looking ahead is indeed part of the business continuity team's responsibilities, to keep an eye on the potential hazards identified in Chapter 2 that may threaten the enterprise at some point. Be confident that you have discerned those hazards well and designed reasonable actions to take when the hazard turns into a disaster. By now, you will have completed a valuable task important to your workplace, employees, and company leaders. This chapter will walk you through the end goal of business continuity planning: organizational resilience or the ability to repel, resist, and bounce back from a disaster's impact. The chapter will also invite you into other vital steps for your enterprise, mitigation planning to reduce future risks and sustainable recovery strategies that promote further resilience. We begin by looking into the future of what businesses may face so that none of our leave our plans on a shelf to gather dust or buried in an electronic file that is rarely retrieved.

Disaster losses and future predictions

Nearly thirty years ago noted disaster researcher Quarantelli (1992) predicted that we would face new, bigger, and more difficult disasters. His scientific forecast has been proven correct, as the world has seen from September 11th, the Indian Ocean tsunami, the Japanese tsunami and nuclear plant accident,

Business Continuity Planning. https://doi.org/10.1016/B978-0-12-813844-1.00010-5

earthquakes in Haiti and Nepal, and hurricanes Katrina and Maria to name a few. Increasing, worsening, and new disasters will cause economic losses without interventions that enhance resilience. One study found that, from 1900 to 2015, approximately 6.5–14.0 trillion U.S. dollars were lost to disasters in amounts adjusted for 2015 (Daniell, Wenzel, & Schaefer, 2016). Businesses cannot continue to sustain such losses without looking ahead to the significant impacts that may occur and what can be done to become more resilient. New threats may also appear that we have not yet discerned. Let us first review a few of the most recently concerning hazards.

Certainly, cybersecurity threats represent one of the most disturbing hazards of the present and the future. From cybercriminals to intentional foreign government intrusion, the potential for catastrophic infiltration represents a significant threat. Worldwide, people rely on Internet capabilities for life-saving cell phone use in remote parts of the world to the management of massive defense systems. Businesses have been targeted for attacks that require payment in electronic currency or have had their intellectual and creative property stolen, which undermines their revenue stream. Reputations have been damaged by cyber intrusions as have election systems and processes. During the COVID-19 epidemic, cybercrimes escalated, disrupting business operations worldwide. Businesses will need to remain vigilant toward cyber threats due to our increasing reliance on technology for managing information, creating and producing products, and communicating.

Vigilance will also be necessary to face continuing threats from active attackers and terrorism. Threat levels should be monitored by business professionals when issued by governments. For example, the U.S. Department of Homeland Security (2020) released its first ever Homeland Threat Assessment in October 2020. These types of information releases make it critical that links are maintained for current information whether those are through industry specific Information Sharing and Analysis Centers (ISACs) or other local contacts. The level of effort that goes on behind the scenes to determine threat level changes is significant. If a threat level increases, it is certainly time to pay attention. Businesses should also think broadly about a terror attack, which could be a direct assault like the world saw on September 11th or a cyber terror attack that takes out a power grid. Terrorism should also be thought of broadly, with potentially either domestic and international origins. The 1995 attack in Oklahoma City that destroyed the Murrah Federal Building and claimed 168 lives, including those in a childcare center, resulted from domestic terror in the U.S. Scale should also be considered, from organization-based terror attacks like the one on the 2015 Westgate Mall in Nairobi, Kenya to the lone wolf actors who assaulted employees gathered for a party in California. Other types of threats should be considered as well, such as those that arise from disgruntled employees or domestic violence. Disgruntled employees can also use other means to disrupt a business including attacks on cyber systems. Direct and indirect impacts could certainly occur along with the potential for downtime and displacement. Those impacts could surface from an immediate threat that shuts down a business to one that is more distant but has indirect impacts.

And, while vigilance toward cyber and terror attacks must be on any business threat radar, extreme weather events remain the most likely hazard that businesses will face. Floods remain the most significant threat worldwide. Climate change is increasing potential exposure to multiple hazards, particularly the worldwide threat of coastal flooding (e.g., see Bangalore, Smith, & Veldkamp, 2019; Ngin, Chhom, & Neef, 2020). Not only are coastal areas at risk, but inland flooding concerns are increasing as well. Recent spatial changes in flooding appear to be initiated by climate change. China, for example, has observed increased precipitation in the northern part of the country while drought has been moving south (Chou, Xian, Dong, & Xu, 2019). Climate change also escalates the potential for drought with related increases in bush/wildfire in Australia and North America (e.g., see Williams et al., 2019). Such conditions carry implicit warnings about food production and distribution as well as the impact of flooding, heat, and water shortages on various industries and international commerce and abilities to transport and receive products in the global supply chain.

Looking into the future can certainly be unsettling. But business continuity planning increases readiness. Coupled with some additional efforts, companies and their employees can prove themselves able to withstand impacts even while facing even the most difficult of crises.

Resilience

Resilience has been defined and measured in a variety of ways (Linnenluecke, 2017; Tierney, 2019). Most generally, resilience is "the ability of an organization to resist, absorb, recover and adapt to the altered environment following a disaster" (Tracey, O'Sullivan, Lane, Guy, & Courtemanche, 2017, p. 1). Business continuity planning promotes resilience through its careful, stepwise effort to address direct and indirect impacts, downtime, and displacement. By identifying critical functions and how a business will handle disruptions to its mission, work teams and planners launch an effort to build resilience for the building, employees, production capacities, supply chains, and customer service. Planning, training, and exercising further deepens organizational abilities to "resist, absorb, recover and adapt" to what disasters cause. If you are at the point of having completed your business continuity plan, you have been engaged in a vital service to your employer and co-workers that offers a more

---

BOX 8.1 Business continuity standards (Last Accessed June 30, 2020).

American Standard: ASIS SPC.1-2009, from the American National Standards Institute, holds organizational resilience at the heart of its goals.

Both the *U.S. based National Fire Protection Association's* NFPA 1600: 2010 and *Canada's* CSA Z1600 emphasize emergency and business continuity management. Content on NFPA 1600 must be purchased from nfpa.org while Canada's CSA Z1600 (based on NFPA 1600) can be bought at csa.ca.

Australia and New Zealand: AS/NZS 5050 focuses on reducing business disruptions for disasters. You can learn more at https://www.standards.govt.nz/news/media-releases/ 2010/jul/new-standard-published-for-managing-disruption-related-risk/.

The *International Standard for Organization* has published ISO 22301:2019 on business continuity. Similar to other protocols, this standard recommends "planning, establishing, implementing, operating, monitoring, reviewing, maintaining and continually improving a documented management system that enables organizations to prepare for, respond to and recover from disruptive incidents" with an emphasis on creating plans suitable to a specific enterprise. This ISO standard can be viewed further at https://www.iso.org/standard/50050.html.

---

promising and robust future in the face of those future disaster events. Do not stop now.

Worldwide, commitment to building resilience has been emerging during the past ten years with an emphasis on sustaining efforts via partnered mitigation initiatives. Several standards have emerged (see Box 8.1) to provide guiding principles, processes, and frameworks. While many remain specific to the nation in which the standard evolved, one global framework recognizes the value of business continuity planning to reduce economic impacts for both the public and private sectors. What underlays this "Sendai Framework" is a commitment to work locally as well as collectively to repel the effects of disasters.

The Sendai Framework

Resilience has become an important goal across many economic sectors and governments and has caught on as an important means to increase abilities to face disaster impacts. The United Nations, for example, has promoted a long-term disaster risk reduction strategy that encourages resilience. Organized most recently in the Sendai Framework (see Box 8.2), this international agreement specifies seven goals to reduce disaster impacts between 2015 and 2030. One goal directly addresses the business community: *reduce direct disaster economic loss in relation to global gross domestic product (GDP) by 2030* (United Nations, 2015). To do so, the internationally agreed upon Sendai Framework recommends:

> Business, professional associations and private sector financial institutions, including financial regulators and accounting bodies, as well as philanthropic foundations, to integrate disaster risk management, including business continuity, into business models and practices through disaster-risk-informed

BOX 8.2 Resources and links, Last Accessed June 30, 2020.

Sendai Framework, https://www.preventionweb.net/files/43291_sendaiframeworkfordrren.pdf.
Resilient New Zealand, https://www.civildefence.govt.nz/cdem-sector/plans-and-strategies/national-disaster-resilience-strategy/ and https://www.civildefence.govt.nz/cdem-sector/plans-and-strategies/national-disaster-resilience-strategy/national-disaster-resilience-strategy-summary-version/.
Public Safety Canada, https://www.publicsafety.gc.ca/cnt/rsrcs/pblctns/mrgncy-mngmnt-strtgy/index-en.aspx.

investments, especially in micro, small and medium-sized enterprises; engage in awareness-raising and training for their employees and customers; engage in and support research and innovation, as well as technological development for disaster risk management; share and disseminate knowledge, practices and non sensitive data; and actively participate, as appropriate and under the guidance of the public sector, in the development of normative frameworks and technical standards that incorporate disaster risk management.

(United Nations (2015, p. 10))

Toward these ends, several national-level efforts have emerged that embrace the Sendai Framework and partner with the private sector. For example, Public Safety Canada is building on the Sendai Framework to promote resilience after devastating floods, wildfires, active attackers, and a horrific train accident. The private sector plays a key role in Canada's whole-of-society approach which aims to move all sectors toward an enhanced ability to bounce back from hazardous impacts. To illustrate, Canada created the 2011 Individual Flood Protection initiative that funded flood mitigation for Manitoba homeowners and businesses. The government's financial investment paid 86% of the funding to elevate or move structures out of flood prone areas or to place dikes that increase area protection. Over 2000 businesses and homeowners joined the initiative and paid the remaining 14%. In a different approach, Prince Edward Island's Emergency Measures Organization collaborated on a hazard identification and risk assessment (HIRA, see Chapters 2 and 3). In what should be a familiar process to readers of this volume, a public-private partnership reviewed the HIRA that included 146 hazards using a scenario format (see Chapter 7). The effort then evolved to include public health facilities and provided support to develop appropriate planning.

Resilient New Zealand, an initiative promoted by the nation's Ministry of Civil Defence and Emergency Management, emphasizes that resilience emanates from community-level preparedness. Their approach recommends that businesses invest in organizational resilience through comprehensive business continuity planning that helps to "reduce and manage the factors that are contributing to your risk" (see citation in Box 8.2). Resilient New Zealand promotes a partnered approach that shares resources to address risks, such as the mitigation measures mentioned above that require individual as well as public and private sector dedication of time, human resources, and financial investment. In both Canada and New Zealand, government, business, and individual partnerships have been leveraged to reduce disasters for all. As the Sendai Framework says: *there is a need for the public and private sectors and civil society organizations, as well as academia and scientific and research institutions, to work more closely together and to create opportunities for collaboration, and for businesses to integrate disaster risk into their management practices.* Both New Zealand and Canada serve as effective examples of how to make that happen.

Embedded within the Sendai Framework and efforts like those promoted by New Zealand and Canada lies the importance and value of mitigation.

Mitigation

As mentioned in Chapters 2 and 3, mitigation represents an important action to take before a disaster occurs to reduce potential losses. Those actions can range from erecting a levee system around a village (structural mitigation) to buying business interruption insurance for a cyberattack (nonstructural mitigation). Something to keep in mind is that a disaster aftermath may be the easiest time to put mitigation measures into place. Planners will now have everyone's attention, because no one will want to go through another disaster with the same impacts. Employers and local governments may be more willing than ever to address future losses by integrating mitigation measures now. Once again, it would be opportune for planners to seize the day when those with resources are willing to expand them in support of risk reduction. Even using recent examples of how mitigation would have reduced losses in another community can catch the attention of those in the local business.

The purpose of mitigation is exactly that, to reduce future losses, conserve the financial bottom line, save lives, and reduce property loss. As explained in Chapter 3, structural mitigation (the built environment, like a floodwall, see Fig. 8.1) may represent the best protections against outside intrusions from an array of hazards like floods, tornadoes, terrorism, or dust storms. While a floodwall may be expensive, the cost of investment will certainly be appreciated when a hospital does not flood and can continue serving the public. Nonstructural mitigation measures, even something as inexpensive as installing smoke detectors, can be just as helpful in alerting employees to stop a potentially catastrophic fire. As an informed business continuity planner, you should now shift to encouraging risk reduction actions that offset losses.

FIGURE 8.1
Tropical Storm Lee, 2011 Binghamton, New York flood wall protects hospital. *FEMA News Photo.*

Mitigation needs a champion

Mitigation tends to be one of the phases in the life cycle of emergency management that needs the most attention but gets the least. People do not like to think about disasters and the costs of some mitigation efforts like river levee systems or blast resistant windows. Yet studies confirm that investments in mitigation pay significant dividends even if the disaster occurs years or decades after installation (Rose et al., 2007). The 2001 Nisqually earthquake that impacted Seattle, Washington, for example, caused less damage than anticipated (Freitag, 2001). One reason is because Project Impact, a Federal Emergency Management Agency program implemented at grass-roots levels, involved 100 large businesses and 500 small businesses. Their earthquake mitigation efforts ranged from retrofitting buildings to computer tie-downs. The school system also participated, which led to minimal damage from the 6.8 earthquake that should have caused significant damage. Interestingly, employees at these workplaces also reported engaging in mitigation efforts at their own homes because of learning how to reduce their risks. Because of their efforts, these employees were far more likely to be able to return to work quickly.

Project Impact involved local leaders and stakeholders with envisioning a different future where people acting now exercised influence over what happened later. Local leaders and everyday people became champions for risk reduction (Meo, Ziebro, & Patton, 2004). A similar Canadian effort, Partners in Protection, inspired local governments to introduce appropriate mitigation after the 2003 Okanagan wildfire (Labossière & McGee, 2017). Their efforts required local champions to support new measures and to partner with area businesses. Unexpected side benefits may accrue. For example, some Canadian mitigation initiatives have introduced new employment opportunities that supported the local economy (Christianson, McGee, & L'Hirondelle, 2012). Partnered investments may prove to be the most beneficial, because mitigation measures that benefit the business sector may require resources beyond what a business holds individually.

Pay now or pay later

Mitigation efforts do require investment, though that investment can range from minimal costs to major expenses. The lesson learned, time and again, is that we must pay now for mitigation or pay later with trying to salvage the business. If we pay later, the costs are likely to be significantly more than we had hoped to endure. Mitigation needs to be thought of as an investment in future safety and business viability. As a next step after completing a business continuity plan, mitigation planning should be pursued. Mitigation efforts can range from relatively inexpensive to those that do require partnered investments.

Consider these mitigation measures, for example, that could make a difference:

- *Free*
  - Communicate with your employees about how to reduce losses in their workspaces and homes, from attaching a bookcase to the wall in an earthquake zone to how to backup data before a cyberattack or prevent phishing intrusions.
  - Create an emergency contact list for phones, email, and in hard copy. Give a laminated copy to employees so that they can account for each other in an emergency. Create another list for the work teams that need to be deployed in a disaster to re-launch critical functions.
  - Distribute preparedness information specific to the workplace and for people's homes, pets, families, and commuting experiences. Free, downloadable guides can be secured at many governmental sites including https://www.ready.gov/ and https://www.ready.gov/business (Last Accessed July 1, 2020).
  - Write a business continuity plan *and* a mitigation plan using employee or volunteer time. Consider involving students and faculty from a nearby university that offers an emergency management program.
  - Conduct annual tabletop exercises written within the firm using the scenarios outlined in this volume.

- *Under $500*
  - Update emergency supplies annually from a first aid kit to the go kits so necessary for working remotely or during a displacement. Some kits may be quite inexpensive while others, like what a restaurant would need to transition to takeout/curbside delivery may require a higher investment and storage capacity.
  - Install fire extinguisher/smoke alarm and security systems at reasonable initial costs with possible monthly costs depending on the size of the business. Replace batteries and test systems on a regular basis. Vendor-based fire suppression systems and contracts can be costly but can save the business and its employees.
  - Train employees on first aid, CPR, and AED which may cost nothing if an internal employee can offer a certified course of instruction. Vendors and agencies (e.g., the Red Cross) also offer in person and online training for live-saving activities at relatively low cost.
  - Pre-place supplies and have a plan in place to distribute personal protective equipment (PPE) as needed. Understandably, the cost could range from low to significant depending on the size and type of the business. Replacement of the PPE could also become significant if supplies are needed during an extended event like a pandemic.
  - Implement a plan to backup computer records which may cost as little as a small thumb-size drive or as complex and costly as offsite backups provided by an external vendor. Straightforward, daily backups launched automatically can keep a business operating with a power failure.
  - Reconnect with the emergency management personnel you contacted in Chapters 2 and 3. Launch a monthly business networking event to stay updated and establish a working relationship.
- *More than $500*
  - Install a safe room where employees can gather when severe weather or an active attacker threatens.
  - Purchase insurance sufficient to cover critical losses which may require hazard-specific insurance, increased policies, or specialized policies like business interruption insurance.
  - Hire an experienced consultant to conduct training and exercising around your business continuity plan.
  - Install floodwalls outside of and barricades inside buildings subject to water intrusion.
  - Place concrete bollards, blast resistant windows, and security barriers to repel an active attacker.
  - Create a regular plan to purchase updated computer equipment that can be mobile along with hotspots and devices that enable employees to connect remotely.

- Store key supplies offsite so that the business may pivot as needed to resupply its critical resources as quickly as possible after a disaster.
- Put in a sprinkler system to reduce fire spread coupled with a means to protect equipment and business resources that may not be water resistant.
- Set aside rainy-day funds sufficient to cover 3–6 months of monthly revenue to offset expenses should disaster occur.

Investing in mitigation might seem like a costly procedure at times. However, think about how people have always adapted to risk reductions. People now wear seat belts and motorcycle helmets, reduced driving speeds, and have joined smoking cessation programs after risk reduction campaigns. With COVID-19, wearing masks has quickly become a worldwide norm coupled with handwashing and social distancing. Employees participate in fire drills and learn how to do basic first aid. Workplaces have instituted warning systems for active attackers, put Internet firewalls into place, and installed sprinkler systems. Businesses buy insurance coverage to protect their assets and include additional coverage from the inventories taken in business impact analysis exercises. Hospitals, government buildings, and schools now have barricades to prevent vehicle intrusion, hire security teams, and move people through electronic scanners. As we learn about risk, and ways to reduce it, workplaces and employees have increasingly embraced risk reduction procedures. Doing so will reduce future impacts and shorten the necessary recovery time, suggesting that investing in mitigation measures represent a solid action to take (see Box 8.3).

The recovery process

Business continuity planning focuses on the aftermath of an event, as employees work to restore critical functions and move forward into a different future than envisioned. This recovery period will likely involve a series of phases that you and your business will move through. The first days and weeks may be very hectic as you pick up the pieces and discern the path forward, even with a completed business continuity plan. After all, realtime testing may reveal new twists and turns not previously anticipated. Think of this time as evolving in a series of phases, from short-term into long-term recovery.

During short-term recovery, you may find yourself in a temporary site or addressing downtime that has reduced operational capacity and the number of employees working alongside. Critical functions will be focused on with attention given to business-saving activities and work teams that implement procedures to safeguard operations. You may be moving people and operations to a new site or supporting employees now handling damage in their own homes. This period will transition in time. Long-term recovery will find you moving back in an original or new space as you expand operational capacity

BOX 8.3 Integrating mitigation measures.

This template, from Chapter 3, now includes columns that introduce the possibilities of structural and non-structural mitigation measures. Each business will need to consider the best ways to reduce risks vis-à-vis the hazards that concern them the most. A few examples are given. To apply to your local context, pull from your HIRA in Chapter 3, then fill in the blanks with what has been or needs to be done with structural and non-structural mitigation measures.

Hazard	Previous history in area	Employee impacts in area	Business impacts	Probability of reoccurrence	Low, medium or high probability event?	Structural mitigation measure	Non-structural mitigation measure
Natural disasters							
Flood	1931 Flood through CBD					Levee Dam Sand bags Floodwall	Flood insurance
Hurricane/ cyclone Flood							
Terrorism, cyberattacks, and active attackers							
Terrorism	Sept 11, 2001	3000 deaths	High	Periodic, unknown	Medium probability, high impact	Concrete bollards	Evacuation plan Training
Active attacker Malware attack							Firewall Phishing education
Pandemic/viral							
COVID-19	2020	23,573 deaths 890,000 ill	Extremely high	Unknown, likely seasonal	High probability, high impact		Social distancing Masks Closures
Influenza	Annual	Varies	Varies	Seasonal	High probability, low to medium impact		
Zika	Annual	Varies	Varies	Seasonal	Low probability, medium to high impact		

Continued

BOX 8.3 Integrating mitigation measures—cont'd

Hazard	Previous history in area	Employee impacts in area	Business impacts	Probability of reoccurrence	Low, medium or high probability event?	Structural mitigation measure	Non-structural mitigation measure
Climate change							
Drought							
Coastal flooding							
Wildfire							
Transportation accidents							
Highway pile-up							
Infrastructure failures							
Bridge failure							
Power failure							
Space weather							
Geomagnetic storm							
Solar flares							
Hazardous materials accidents							
Offshore oil spill							
Internal spill/explosion							

toward normal. Realize that this phased recovery process will take time as it allows you, your teams, and your business to repair, rebuild, restore, and return to critical and routine business functions.

The recovery process may also take time depending on the magnitude, scale, scope, and location of the disaster and what needs to be restored. The 2004 tsunami decimated the commercial sector in Vailankanni, India. But with outside support, vendors regrouped quickly and reopened within months. Businesses in the World Trade Center required more time to heal from the traumatic event and to find new spaces from which to operate. Businesses affected by the global pandemic of 2020 will likely ride a wave of transitions with restarting operations paused periodically as hotspots and outbreaks continue until a vaccine becomes available and sufficiently efficacious.

Throughout such challenges, it will become important to adapt, pivot, transition, and change, and to rely on employees, to demonstrate flexibility, and to focus on sustaining your business for future generations. A few insights about recovery conclude this volume starting with knowing that co-workers will be there for you and your business.

Employee support

Many people have assumed that workers will abandon their jobs when disaster occurs. Role abandonment, the assumption that people will not do their jobs in a disaster, has turned out to be a well-documented myth about human behavior (e.g., see Quarantelli, 2008; Trainor & Barsky, 2011). Research shows that you can trust your colleagues. Why? Because people raise their children to behave in prosocial ways and to be helpful to others. Because we train people how to react in an emergency so that they know what to do. Because social expectations compel people to be kind toward each other. People stay on the job and step up when the unimaginable happens.

When the planes hit the World Trade Center, first responders rushed to the scene and hundreds of them gave their lives trying to save others. Co-workers carried colleagues down dozens of flights of stairs and assisted those who were injured. Some willingly stayed behind with their injured or trapped co-workers and lost their lives. As the Indian Ocean tsunami rushed ashore in thirteen nations, people rescued those around them including complete strangers. They cared for each other, offering first aid and emotional comfort in the hours and days they awaited rescue. In every terror attack, colleagues, bystanders, and complete strangers have cared for those around them by sheltering others, tending to injuries, and disabling attackers. The fire mentioned in Chapter 1, at the Beverley Hills Supper Club, resulted in workers saving customers and then helping to direct rescue efforts (Johnston & Johnson, 1989). Even neighbors will be among the first responders to help when disaster occurs, ahead of the police and fire normally tasked with such rescue work. When the coronavirus pandemic spread worldwide, health care workers stayed on the job as did grocery store workers, delivery people, educators, and emergency managers.

Know that when disaster strikes, people step up. Employees will turn out to do their jobs, support the business and each other. Employees will not be the only people to show up and try to help. Indeed, massive volunteer turnouts occur after disasters as people volunteer to clean up the homes and businesses of those who have been harmed (Phillips, 2020). People will arrive spontaneously to help, compelled by the images they see on various media, putting aside their own jobs to travel hundreds of miles to assist. Dozens of volunteer units will also arrive, offering organized work teams to do everything from collecting stray farm animals to sorting debris at a devastated site. People perform heroically, give their time and money to others, volunteer in unpleasant and difficult circumstances, and help launch recovery. Often described as the therapeutic community, the arrival of volunteers will enable a devastated community to turn toward recovery (Barton, 1969; Phillips, 2013). You may need that kind of help, because disasters can present some unusual twists and turns in what you had planned to occur.

Flexible implementation

Even the best plans will require modification when events unfold in realtime because things may not go as intended. A key work team member may be on business travel or injured by the disaster (which is why you have identified backups). You may have just onboarded a new work team member who had not yet been trained on the plan or does not yet know business operations and co-workers well. The disaster may have happened when everyone was off shift, making it difficult to get to the business in a timely manner or through damaged transportation arteries. Someone will have forgotten key parts of their go kits or the computer devices they took home will not operate correctly or may break. The plan to telecommute may have gone awry due to power outages or Internet disruptions. Traditional supply chains may have been significantly disrupted beyond your expectations and planning.

Despite decades of pandemic planning, COVID-19 overwhelmed health care systems and testing facilities, shut down businesses and schools, and generated a significant amount of creative thinking. People learned to pivot as a new, continuing way of life and to innovate their ways through unexpected twists and turns, even in businesses where planning had been worked on for decades. Medical and personal protective equipment shortages led businesses to help create new ways to produce and clean ventilators and masks. Such shortages had not been anticipated even in nations with extensive pandemic stockpiles. As the pandemic ensued, additional challenges appeared, such as an unexpected coin shortage in the U.S. creative businesses managed the coinage scarcity by offering gift cards rather than change, posting additional points on shopper reward cards, and accepting only electronic payments. When the unimaginable happens and plans fail to account for all conditions, it will be necessary to demonstrate resilience through creative innovation (Kendra & Wachtendorf, 2007).

Expect that things will happen and that disasters will challenge our most certain assumptions and well-exercised procedures. Consider a comparable example outside of traditional disasters. Pilots who test out new airplanes learn the flight procedures and plans in exacting detail. Nonetheless, things go wrong that the plan and prior training do not cover. Innovating and thinking creatively is what keeps those pilots alive, such as Brigadier General Chuck Yeager, the first pilot to break the sound barrier despite some challenging flat spins and crashes (Wolfe, 1979). Indeed, research has confirmed that planning, coupled with an ability to be flexible and to innovate, promotes post disaster and economic resilience (Graveline & Gremont, 2017; Orchiston, Prayag, & Brown, 2016). Be flexible and plan to be creative (Webb & Chevreau, 2006).

The training process that you put your BCP team through should have increased their abilities to communicate and collaborate and to discern and troubleshoot innovatively. Thus, planners, work teams, personnel, and leaders should all embrace the idea that flexibility and adaptation represent ideal behaviors and encourage employees to face the unknown creatively. Have confidence in your team's ability to execute critical functions and to roll with the challenges that disasters present. Your company's biggest investment will have been in the people because plans cannot innovate. People innovate and people *are* the plan.

Sustainable recovery

When people step up, employers can reward their workers' efforts by investing recovery funds into initiatives that increase future resilience. The mitigation measures encouraged throughout this volume represent one path to do so. Increasingly, another creative post-disaster rebuilding strategy has incorporated approaches that embrace sustainability.

For example, communities are increasingly introducing green infrastructure when opportunities to repair or rebuild a disaster-damaged area occurs. As one installable strategy, permeable surfaces and green spaces can reduce stormwater runoff and related flooding. Streetscaping can repel stormwater through increasing tree canopy while also improving air quality and can reduce the effects of landslides and mudslides (Berland et al., 2017; Nowak & Dwyer, 2007).

Native plant wildscapes or xeriscapes can be introduced to reduce hazards from drought and wildfire threats. Such urban islands also support insects beneficial to the environment (Boger & Marr, 2018).

Another increasingly used option is to pivot toward zero energy building (Marszal et al., 2011). For example, repairing or rebuilding may introduce energy efficiency measures such as upgraded lighting and HVAC systems, low energy computer monitors, and low-flow faucets. Businesses might also pursue LEED certification by following specific criteria to reach various recognition levels that demonstrate energy efficiency: certified, silver, gold, and platinum status (see https://www.usgbc.org, Last Accessed June 30, 2020). Solar panels can reduce reliance on the power grid. Recycled materials can be used.

For more specific hazards, new materials may be introduced, such as placing hurricane clamps on roofs to repel high wind events installing safe rooms for tornadoes. Businesses may want to purchase and place generators for possible power outages or integrate blast resistant windows for terror attacks, active attackers, and tornadoes. Businesses could work with utility companies to place power lines underground or to replace a deteriorated part of the broader infrastructure. Businesses might also elevate their structures to reduce future water intrusion from a hurricane or the expected consequences of climate change. Health care plans can expand to include immunizations and coverage for protective equipment that reduces the spread of an outbreak. Recovery represents an opportunity to build back not only better but stronger.

In short, recovery can be used to reduce future threats, support the community and its environment, and save on future expenses. Business continuity planning, mitigation measures, and recovery efforts all work together to reduce future risks from an array of hazards. Appropriately then, our goal in writing this book has been to provide a stepwise guide to business continuity planning and to invite planners to work alongside the emergency management community in reducing risks, preserving lives and property, and building a more sustainable and resilient future for all. We invite you to remain engaged in protecting your business and employees.

Essential actions

- Keep training, exercising, revising on your plan.
- Update your business continuity plan on a regular basis, at least annually.
- Remain positive because you can influence your company's future and coworker livelihoods.
- Practice creative thinking and be innovative when presented with recovery challenges.

- Participate in and support mitigation planning across your community and region, as it will reduce your business risks as well.
- Create a mitigation plan for your business or agency.
- Invest in mitigation measures that reduce risks and the time needed to recover.
- Introduce post-disaster initiatives that promote sustainability, reduce reliance on natural resources, and reduce future risks.
- Be the champion your workplace needs.

References

Bangalore, M., Smith, A., & Veldkamp, T. (2019). Exposure to floods, climate change, and poverty in Vietnam. *Economics of Disasters and Climate Change, 3*(1), 79–99.

Barton, A. (1969). *Communities in disaster: A sociological analysis of collective stress situations.* NY: Doubleday.

Berland, A., Shiflett, S. A., Shuster, W. D., Garmestani, A. S., Goddard, H. C., Herrmann, D. L., & Hopton, M. E. (2017). The role of trees in urban stormwater management. *Landscape and Urban Planning, 162,* 167–177.

Boger, A. M., & Marr, D. L. (2018). Effect of native and non-native plantings in urban parking lot islands on diversity and abundance of birds, arthropods, and flower visitors. *Proceedings of the Indiana Academy of Science, 127*(3), 115–123.

Chou, J., Xian, T., Dong, W., & Xu, Y. (2019). Regional temporal and spatial trends in drought and flood disasters in China and assessment of economic losses in recent years. *Sustainability, 11*(1), 55.

Christianson, A., McGee, T. K., & L'Hirondelle, L. (2012). Community support for wildfire mitigation at Peavine Métis Settlement, Alberta, Canada. *Environmental Hazards, 11*(3), 177–193.

Daniell, J., Wenzel, F., & Schaefer, A. (2016). The economic costs of natural disasters globally from 1900-2015: Historical and normalised floods, storms, earthquakes, volcanoes, bushfires, drought and other disasters. In: *EGUGA, EPSC2016-1899.*

Freitag, R. (2001). The impact of Project Impact on the Nisqually earthquake. *Natural Hazards Observer, 25,* 5.

Graveline, N., & Gremont, M. (2017). Measuring and understanding the microeconomic resilience of businesses to lifeline service interruptions due to natural disasters. *International Journal of Disaster Risk Reduction, 24,* 526–538.

Johnston, D. M., & Johnson, N. R. (1989). Role extension in disaster: Employee behavior at the Beverly Hills Supper Club fire. *Sociological Focus, 22*(1), 39–51.

Kendra, J., & Wachtendorf, T. (2007). Improvisation, creativity, and the art of emergency management. *Understanding and responding to terrorism* (pp. 324–335)Vol. 19, (pp. 324–335). .

Labossière, L. M., & McGee, T. K. (2017). Innovative wildfire mitigation by municipal governments: Two case studies in Western Canada. *International Journal of Disaster Risk Reduction, 22,* 204–210.

Linnenluecke, M. K. (2017). Resilience in business and management research: A review of influential publications and a research agenda. *International Journal of Management Reviews, 19*(1), 4–30.

Marszal, A. J., Heiselberg, P., Bourrelle, J. S., Musall, E., Voss, K., Sartori, I., & Napolitano, A. (2011). Zero energy building—A review of definitions and calculation methodologies. *Energy and Buildings, 43*(4), 971–979.

Meo, M., Ziebro, B., & Patton, A. (2004). Tulsa turnaround: From disaster to sustainability. *Natural Hazards Review, 5*(1), 1–9.

Ngin, C., Chhom, C., & Neef, A. (2020). Climate change impacts and disaster resilience among micro businesses in the tourism and hospitality sector: The case of Kratie, Cambodia. *Environmental Research,*109557.

Nowak, D. J., & Dwyer, J. F. (2007). Understanding the benefits and costs of urban forest ecosystems. In *Urban and community forestry in the northeast* (pp. 25–46). Dordrecht: Springer.

Orchiston, C., Prayag, G., & Brown, C. (2016). Organizational resilience in the tourism sector. *Annals of Tourism Research, 56*, 145–148.

Phillips, B. (2013). *Mennonite Disaster Service: Building a therapeutic community after the gulf coast storms.* Lexington Books.

Phillips, B. (2020). *Disaster volunteers: Recruiting and managing people who want to help.* Cambridge, MA: Elsevier Press.

Quarantelli, E. L. (1992). *The environmental disasters of the future will be more and worse but the prospects is not hopeless. Available at (1992).* http://dspace.udel.edu/bitstream/handle/19716/574/PP186.pdf?sequence=3&isAllowed=y. *(Last Accessed June 30, 2020).*

Quarantelli, E. L. (2008). Conventional beliefs and counterintuitive realities. *Social Research: An International Quarterly, 75*(3), 873–904.

Rose, A., Porter, K., Dash, N., Bouabid, J., Huyck, C., Whitehead, J., & Tobin, L. T. (2007). Benefit-cost analysis of FEMA hazard mitigation grants. *Natural Hazards Review, 8*(4), 97–111.

Tierney, K. (2019). *Disasters: A sociological approach.* Cambridge, UK: Polity Press.

Tracey, S., O'Sullivan, T. L., Lane, D. E., Guy, E., & Courtemanche, J. (2017). Promoting resilience using an asset-based approach to business continuity planning. *SAGE Open, 7*(2) 2158244017706712.

Trainor, J., & Barsky, L. (2011). *Reporting for duty? A Synthesis of research on role conflict, strain, and abandonment among emergency responders during disasters and catastrophes.* Disaster Research Center.

United Nations (2015). *Sendai framework for disaster risk reduction 2015–2030.* (2015). https://www.preventionweb.net/files/43291_sendaiframeworkfordrren.pdf.

U.S. Department of Homeland Security (2020). *Homeland threat assessmnet.* (2020). https://www.dhs.gov/sites/default/files/publications/2020_10_06_homeland-threat-assessment.pdf. Accessed 10 October 2020.

Webb, G. R., & Chevreau, F. R. (2006). Planning to improvise: The importance of creativity and flexibility in crisis response. *International Journal of Emergency Management, 3*(1), 66–72.

Williams, A. P., Abatzoglou, J. T., Gershunov, A., Guzman-Morales, J., Bishop, D. A., Balch, J. K., & Lettenmaier, D. P. (2019). Observed impacts of anthropogenic climate change on wildfire in California. *Earth's Future, 7*(8), 892–910.

Wolfe, T. (1979). *The right stuff.* New York: Bantam.

Appendix 1. Emergency response planning basics

While writing a business continuity plan, many questions will arise about the emergency response time and how the business is working on life-saving matters. Research indicates that many businesses do the minimum when it comes to emergency situations, especially large disasters. Further, businesses do not always update what they have done including first aid kits and training. Some industries must subscribe to standards and policies that influence their workplaces, such as those created by various occupational and safety groups. Other businesses will need to address requirements set out by government and agency mandates, such as environmental safety standards that specify how to manage chemicals and protect workers.

A business emergency response plan should address, as a minimum, these matters:

- Contact information for fire, police, ambulance, emergency medical teams, hospitals, and 9-1-1 types of emergency telecommunications.
- Instructions on how to report various types of emergencies (e.g., fire, bomb threat, tornado).
- Location and use of emergency equipment, from basic first aid kits to AED units, eyewash stations, and emergency evacuation resources (e.g., for employees with mobility challenges).
- Ways to alert employees to impending hazards from a rapid onset event like an explosion to a longer-term event including pandemics or droughts. Strategies should cover a range of ways to communicate from loudspeakers to email, texting, weather radios, cell alert systems, and specialized alert devices. Care should be given to introduce means to alert employees, customers, patrons, patients, and others who may be on the business property including ways to inform people who may experience hearing challenges.
- Protocols for specific hazards, such as the rapid onset of a tornado that may require sheltering in a place of safety. Similarly, a chemical accident could introduce vapors into a business area that require sheltering in

place differently including shutting down air conditioning systems. An active attacker may require a range of strategies from sheltering in place to evacuating when possible.

- Procedures for evacuating from an affected area to a location of safety. Specific circumstances should be considered from a tornado to an active attacker to determine best practices.
- Protective equipment for area hazards which can range from safety glasses to high level protective suits, oxygen tanks, and shelters from fire, tornado, floods, and other hazards.
- Directions for where to go and how to get there.
- Training and educating employees on how to react to specific kinds of hazards and practicing that behavior. Practical training can and should include basic first aid, CPR/AED, fire extinguisher use, the Heimlich maneuver, and tourniquet use. Training should also include how to de-escalate potentially volatile situations and when to take shelter from a human threat.
- Accountability measures and communication strategies for how to contact employees to confirm they are safe.
- Contact list for supervisors with a call-down procedure, phone tree, or group texting procedure.
- Encourage emergency preparedness at home and while traveling as further measures to safeguard your workplace and its most valuable assets, employees.

Resources (last accessed July 23, 2020):

U.S. OSHA requirements, Occupational Safety and Health Administration (OSHA) regulation 1910.38(b), https://www.osha.gov/laws-regs/regulations/standardnumber/1910/1910.38.
Basic emergency plan, https://www.ready.gov/business/implementation/emergency.
First aid and related classes, https://www.redcross.org/take-a-class.
Federal Emergency Management Agency, IS-235c Emergency Planning, https://training.fema.gov/is/courseoverview.aspx?code=IS-235.c.
U.S. Department of Homeland Security, https://www.dhs.gov/how-do-i/prepare-my-business-emergency.
U.S. Department of Homeland Security, National Incident Management System, http://www.fema.gov/media-library/assets/documents/148019.
Hazard specific information, https://www.ready.gov/.
Basic disaster supplies kit, https://www.ready.gov/kit.
Vehicle emergency kit, https://www.ready.gov/car.

Appendix 2. Mitigation planning basics

Mitigation planning typically unfolds as a community-wide enterprise, usually led by a local emergency management agency. That EMA invites stakeholders into the planning process and walks everyone through discerning hazards, assessing risks, and prioritizing recommendations that reduce those risks. Planners who have completed this volume would be in high demand to participate on a mitigation planning effort for a community because of their familiarity with the HIRA process described in Chapter 3. In addition, businesses would be wise to support community-based mitigation measures that reduce risks for all including the enterprises active in the area. Another good reason to participate is because the costs of risk reduction will be shared across the community and reduce the economic burden for all. Think of participating in a mitigation planning effort as another form of investing in insurance, but this time the investment comes from your time well spent.

In addition to the community-wide efforts that introduce and seek funding to retrofit buildings, strengthen levee systems, and protect infrastructure, businesses should also take the time to do what they can to mitigate risks. That internal effort might look like this:

Convene the Business Continuity Planning Team. Discuss the importance and value of mitigation with the team and see who would be willing to continue on an internal mitigation team and who might be willing to serve as a liaison to any broader community efforts. Discuss additional new team members who might be brought on board to enrich the discussion and further involve employees in risk reduction efforts.

Inform the team. Provide your mitigation planning team with the information they need to succeed. They can take free courses from the FEMA independent study course series available at https://training.fema.gov/is/. Courses that should be of interest include:

S-318 Mitigation Planning for Local and Tribal Communities
IS-319.a Tornado Mitigation Basics for Mitigation Staff
IS-320 Wildfire Mitigation Basics for Mitigation Staff
IS-321 Hurricane Mitigation Basics for Mitigation Staff
IS-322 Flood Mitigation Basics for Mitigation Staff

Two additional free programs funded by the Department of Homeland Security at Texas A&M University – Texas Engineering Extension Service can assist in providing background to mitigation planning for cyber events and infrastructure security and resilience. https://teex.org/program/cybersecurity/

1. Cybersecurity online training for Business Professionals (Cyber 301) track
   AWR169 – Cyber Incident Analysis and Response
   AWR176 – Disaster Recovery for Information Systems
   AWR177 – Information Risk Management
2. Infrastructure Protection Certificate Program
   https://teex.org/program/infrastructure-protection/
   AWR213 – Critical Infrastructure Security and Resilience Awareness
   MGT310 – Threat and Hazard Identification and Risk Assessment and Stakeholder Preparedness Review
   MGT315 – Critical Asset Risk Management
   MGT414 – Advanced Critical Infrastructure Protection

Prioritize hazards. Using the Template from Chapter 8, rank the hazards identified in your HIRA (from Chapter 2) in terms of how much they concern you and threaten the business. Be sure to fill out what you have already implemented as mitigation measures, from weather radios to insurance and employee education efforts.

Determine what needs protection. Focus on the assets of the business including the people, production capacities, buildings, vehicles, and other resources. Review the inventories that you did for your business continuity plan. In what ways might these resources and assets be at risk in the hazards of most concern?

Contact your emergency management agency. What existing mitigation measures are in place to protect your business? Does the jurisdictions have a current Hazard Mitigation Plan? What are being worked on? Do you need additional mitigation measures or should you join a community effort to fund them through other means?

Surface new mitigation measures. Through additional discussion and research, identify additional mitigation measures that could effectively increase resilience with a future event of concern. This will take research which can include tips and ideas from the resources listed below.

Decide. Which mitigation measures represent the best fit between hazards of concern and the company's ability to invest? Select the mitigation measures that work best.

Invest. Decide to make the investment that works best for you and your company. All businesses and jurisdictions must make judgment calls about what will be the best route they can take. You may not be able to do all you want, but you can do something.

Resources, last accessed July 23, 2020:

Mitigation measures, including those under $50, https://www.fema.gov/mitigation-ideas-and-tips-rebuilding and for businesses: https://www.fema.gov/getting-down-business.
FEMA Terrorism mitigation information for attacks on buildings, https://www.fema.gov/media-library/assets/documents/2150.
FEMA, *Mitigation Ideas: A Resource for Reducing Risk to Natural Hazards* https://www.fema.gov/media-library-data/20130726-1904-25045-0186/fema_mitigation_ideas_final508.pdf.
Faith-based resources for mitigation, https://www.fema.gov/faith-resources.
FEMA's mitigation video toolkit, https://www.youtube.com/watch?v=mmAsy3PbYes.

Worth the time to read:

Chikoto, G. L., Sadiq, A. A., & Fordyce, E. (2013). Disaster mitigation and preparedness: Comparison of nonprofit, public, and private organizations. *Nonprofit and Voluntary Sector Quarterly, 42*(2), 391–410.
Prayag, G., & Orchiston, C. (2016). Earthquake impacts, mitigation, and organisational resilience of business sectors in Canterbury. In *Business and post-disaster management: Business, organisational and consumer resilience and the Christchurch earthquakes* (pp. 97–120).
Yoshida, K., & Deyle, R. E. (2005). Determinants of small business hazard mitigation. *Natural Hazards Review, 6*(1), 1–12.

Appendix 3. Essential tools and resources for business continuity planning

This volume includes examples of multiple templates that introduce how planners how to move through the planning process. Readers are welcome to contact the authors to receive original word documents of these forms and templates.

Many nations and organizations have introduced the idea of business continuity planning to become more resilient. Toward that end, they have posted numerous tools, templates, handouts, and procedures on various websites. Certainly one of the most robust websites is the U.S. based Federal Emergency Management Agency's Business Continuity Planning Suite at https://www.ready.gov/business-continuity-planning-suite. The suite includes templates, handouts, videos, worksheets, toolkits, guides, and a software program that generates a business continuity plan. Care should be taken to look at the materials and discern their value for a business. Some materials may be more than sufficient for a business while others may only be a start. The difference, for example, between a local seamstress and a nuclear plant demonstrates the challenge of creating a single set of materials. For the seamstress, the most straightforward and shorter materials may be sufficient, such as templates provided at the FEMA site. A nuclear power plant will need to develop a significantly larger and more detailed plan that will need to address many policies and procedures required from a range of agencies and government bodies.

In addition, the U.S. Department of Homeland Security maintains a free training program – Continuity Excellence Series https://www.fema.gov/continuity-excellence-series-professional-and-master-practitioner-continuity-certificate-programs.

The series includes the following FEMA Independent Study online courses:

IS 546.a: COOP Awareness Course
IS 547.a: Introduction to COOP

IS 1300: Introduction to Continuity of Operations
IS 545: Reconstitution Planning Workshop
IS 520: Introduction to Continuity of Operations Planning for Pandemic
Influenza Course

Index

Note: Page numbers followed by *f* indicate figures and *b* indicate boxes.